Mastering

Physics

Palgrave Master Series

Accounting
Accounting Skills
Advanced English Language
Advanced Pure Mathematics
Arabic
Basic Management
Biology
British Politics
Business Communication
Business Environment
C Programming
C++ Programming
Chemistry
COBOL Programming
Communication
Computing
Counselling Skills
Counselling Theory
Customer Relations
Database Design
Delphi Programming
Desktop Publishing
e-Business
Economic and Social History
Economics
Electrical Engineering
Electronics
Employee Development
English Grammar
English Language
English Literature
Fashion Buying and Merchandising
  Management
Fashion Marketing
Fashion Styling
Financial Management
Geography
Global Information Systems

Globalization of Business
Human Resource Management
Information Technology
International Trade
Internet
Java
Language of Literature
Management Skills
Marketing Management
Mathematics
Microsoft Office
Microsoft Windows, Novell
  NetWare and UNIX
Modern British History
Modern European History
Modern United States History
Modern World History
Networks
Novels of Jane Austen
Organisational Behaviour
Pascal and Delphi Programming
Philosophy
Physics
Poetry
Practical Criticism
Psychology
Public Relations
Shakespeare
Social Welfare
Sociology
Spanish
Statistics
Strategic Management
Systems Analysis and Design
Team Leadership
Theology
Twentieth-Century Russian History
Visual Basic
World Religions

www.palgravemasterseries.com

Palgrave Master Series
Series Standing Order ISBN 0–333–69343–4
(outside North America only)

You can receive future titles in this series as they are published by placing a standing order. Please contact your bookseller or, in case of difficulty, write to us at the address below with your name and address, the title of the series and the ISBN quoted above.

Customer Services Department, Macmillan Distribution Ltd
Houndmills, Basingstoke, Hampshire RG21 6XS, England

Mastering

Physics

Fourth Edition

Martin Harrison
Frank McKim

Founding authors: John Keighley and
Frank McKim

MACMILLAN

First published 1999 by
MACMILLAN PRESS LTD
Houndmills, Basingstoke, Hampshire RG21 6XS
and London
Companies and representatives throughout the world

ISBN-10: 0- 333- 69874- 6 paperback
ISBN-13: 978-0- 333- 69874- 7 paperback

A catalogue record for this book is available from the British Library.

This book is printed on paper suitable for recycling and made from
fully managed and sustained forest sources.

10   9   8   7   6   5   4   3
07   06

Printed in China

Contents

Preface to the first edition

The book aims to provide a concise, easily readable treatment of all the essential principles contained in introductory physics courses. We have tried to present them with a directness and simplicity that will enable students to achieve maximum comprehension in the shortest possible time. Many diagrams have been included in the text as these are a great help in understanding physics and are especially useful in revision.

MARLBOROUGH, 1982

JOHN KEIGHLEY
FRANK MCKIM
ALAN CLARK
MARTIN HARRISON

Preface to the fourth edition

This new edition has been prompted both by changes in syllabus and in ways of assessing a student's abilities in physics. It has meant putting together a new book rather than updating the previous edition. Most of the illustrations are new, having been chosen to show how physics is important in the natural world, as well as in such fields as athletics, engineering, medicine and music. We have also included examples in the text where a grasp of physics has become increasingly important to environmentalists and politicians, as well as in other fields like motor racing and space flight.

MARLBOROUGH, 1999

MARTIN HARRISON
FRANK MCKIM

Acknowledgements

The early editions of Mastering Physics were the work of four collaborators. Dr Alan Clark wrote the final quarter of that book, and we want to pay tribute to his clarity of presentation, setting a standard for his successors to strive for.

John Keighley was the coordinator of the first three editions, and it was due to his experience of authorship, enthusiasm and attention to detail that the original text took shape.

Both men have been our teaching colleagues in the Physics Department at Marlborough College. Our thanks go to that Department and to the Music Department under Robin Nelson, some of whose Symphony Orchestra players are pictured with their wind instruments in chapter 14. They include Candida Sopwith, Hannah O'Regan, Edward Baring, Edward Cooke, Robert Greville-Heygate, Basil King, and Henry Jeens.

Our thanks are also due to the following who have kindly provided illustrations and information for our examples: Richard Austin, Rebecca Bangay at the BBC Natural History Film Library; Graham Bell; Sylvia Chaplin at the Nuffield Radio Astronomy Laboratories, Jodrell Bank; Kevin Conkey, Chris Freemantle at the Wellcome Department of Cognitive Neurology; Anne Froggatt at the Blackpool Pleasure Beach; Frank Greenaway; Charles Healey; Haydn Jones; Nick Parks at Marlborough College Outdoor Activities Department; Mark Shearman; Bruce Tulloh; California Institute of Technology; Civil Aviation Authority; Falcon Cycles Ltd; Mount Wilson and Palomar Observatories; NASA; Projects and Modules Team at

Maplin Electronics plc; Rolls-Royce plc; the Science Photo Library; the Transport Research Laboratory; and the Wellcome Department of Cognitive Neurology.

We are grateful to both the London and Northern Ireland Examination Boards for permission to reprint GCSE questions from their past papers.

Introduction

The content of this book has been determined by the GCSE physics syllabus in the United Kingdom, which is comparable with first level High School physics in some other countries. Our aim has been to describe the physical properties of materials, and the laws of physics relating to observations we can make both inside and outside the laboratory.

We have included a limited number of mathematical examples, encouraged by the thought that Michael Faraday, who has been described as the greatest experimental physicist of the nineteenth century, used no equations at all in his papers. In this book the numerical work is no more complicated than most physics courses at this level require. However, there are equations throughout the book. We hope you will find that they provide a concise summary of relationships between physical quantities, and can be simply applied.

We have tried to keep the language straightforward, explaining new technical terms where they arise in the text. Words used in a scientific sense sometimes need careful definition and understanding. For example 'power', 'energy', 'momentum', 'force', 'resistance' and 'moderator' all have non-scientific meanings. To use words like these accurately in context is to make your reasoning clear.

Many physical quantities can be given numerical values. These have meaning only if a system of units is defined for each quantity. We shall use units based on the metre of length, the kilogram of mass and the second of time. This is called an m.k.s. system of units, and it is often referred to as the SI system (Système International). Multiples and submultiples of these units are also used for convenience (see Table 1 below).

So the following are all equal amounts of energy: 1 000 000 J, 10^6 J, 1 MJ, and 1000 kJ.

We do not list all the experiments you might find in a typical course book. However, there are accounts of experiments which help to illustrate physical phenomena. As you

Table 1 Some common prefixes with their meanings and abbreviations

prefix	meaning	symbol	
giga-	$\times 10^9$	G	The universe may be older than 10 gigayears
mega-	$\times 10^6$	M	A power station may deliver 500 megawatts
kilo-	$\times 10^3$	k	A mile is about 1.6 kilometres, i.e. 1.6×1000 m
centi-	$\times 10^{-2}$	c	A wine bottle holds about 75 centilitres, or (75/100) li
milli-	$\times 10^{-3}$	m	This page is about 0.1 millimetre thick, or (0.1/1000) m
micro-	$\times 10^{-6}$	μ	A soap film may be a few micrometres thick

read through a chapter you may find a self test question in the text, together with an answer you should be able to obtain. The previous part of the text should be reinforced by your answering the question. Further questions based on the text occur at the ends of chapters, and answers to some of these may be found at Appendix D. In Appendix C you will find some revision questions which have been asked in GCSE examinations in the UK.

This book could be used for revision of the subject. However, for those who use it to accompany their coursework, some hints on planning and carrying out a practical physics project may also be helpful. We use the rest of this introduction to consider how you might plan, carry out and write up a record of your work.

Experimental project work

Several physics courses devote some time to an assessed practical project. Typical subjects for study might be:

1   the time it takes a simple pendulum to swing forwards and backwards again
2   the greatest friction force which occurs when you try to pull a wooden block along a flat surface. (This is the force you must apply to get the block to slide)
3   the energy transferred to solids or liquids from a current in an electric heating element
4   the rate of loss of heat from a can of hot water surrounded by a layer of lagging.

Whatever the assignment may be, it will be helpful to have an outline of the stages of planning, conducting experiments, analysing results and concluding your report. The following outline shows the steps which should be recorded in the write-up of your work.

Practical assignment outline

1   Planning
2   Conducting experiments
3   Analysing results and concluding

---

1   **Planning** (you keep a record of each of these steps)
   a   **Identify** and make a list of the possible
       variables. These are the things you could change
       in order to investigate their effect on the
       phenomenon you are given to study. With a
       pendulum study you might choose its length and
       the mass of the bob as variables, or you might
       think of others.
   b   Devise experiments **adjusting one variable at a
       time**.
   c   Make a list of the apparatus you will use.
   d   Show how the apparatus is to be set up for each
       test.

e   Decide how you will obtain the readings in each test.
f   Show how you will record your readings.
g   Predict what you expect your results to show.

2   **Conducting experiments** (to obtain and record readings **with units**)
a   Set up the apparatus and try some simple tests.
b   See if you can obtain repeatable readings.
c   Conduct each of the planned tests and **record readings clearly**.
    (Repeated readings may lead to reliable average results).

3   **Analysing results and concluding your write-up**
a   Plot graphs and complete calculations.
b   Use results to assess the effect of each variable.
c   Express your findings in conclusions, which express general trends and/or more precise mathematical relationships.
d   Now have a look at 1 g above and compare your findings with your predictions.
e   Say what improvements you could make to the experiments.

A practical scientist records numerical data to an appropriate number of significant figures, and you can be marked down for giving answers to calculations to more figures than your data can justify. A good rule of thumb is not to use more than three sig. figs, and be ready to round up your answers to two sig. figs unless you have very accurate instruments and very repeatable results. So, for **3.86 mm** you write **3.9 mm** (2 sig. figs), but for **3.84 mm** you write **3.8 mm** (2 sig. figs).

On completion of your project you should check that your name is on each page. The front page should carry a title, e.g. '**The swings of a pendulum. Pete Jones. Set P.**' Check that your graphs are properly plotted on squared paper with the axes clearly ruled, numbered, and labelled. The graphs also need headings, e.g. '**Graph of time for ten swings of a pendulum plotted against the pendulum's length**'. Number the pages in order, and clip or staple them together. Anyone who reads your report should then be able to grasp what it was you studied, how you set about it, what you found out and how reliable you think your results are.

1 ❯ Handling materials

Objectives

After this chapter you should
- know the units of density and understand how density can be measured
- be able to calculate a density from mass and volume data
- know how the weight of an object on Earth is related to its mass
- understand that objects weighing down on the Earth's surface exert a pressure
- be able to relate the pressure on a surface to the surface area and the force applied to it
- understand what influences the pressure exerted at the foot of a column of liquid
- know that the pressure at a point in a fluid is exerted equally in all directions.

1.1 Introduction

Play-Doh® is rather like plasticene. If you wanted to show that these two substances are different, how would you set about it?

1.2 Data from experiments – Plan A

You could start by looking at a lump of each material to examine its texture. Then you might cut each lump in half and examine the sections to see if the materials have the same appearance inside as outside.

Perhaps you would use your sense of smell, as plasticene usually has a giveaway smell. Alternatively you might see how easy it is to roll each lump of material in your hands until it becomes sausage-shaped. You may detect a difference between the materials; but it would be best to continue working the pieces in your hands as this will warm them up. Do they both become easier to work as they warm up?

Try prodding each lump of material with your finger and compare the impressions in each sample. If your finger will not sink in, repeat this test with a thumb tack or drawing pin and see which material is better at resisting the pinpoint.

These are only some of the tests you might carry out. For example, you could also try a stretching or twisting test.

As a scientist you need to **keep a record of the results** of your tests. You may convince yourself that the lumps are of different materials but a written record, made when you do the tests, can be referred to later when you need to provide the evidence.

1.3 Obtaining further data – Plan B

You could find out if the two materials have different densities. The density of a substance, (in symbols d or ρ), is the amount of matter it contains in each unit of its volume. **Density is the ratio of total mass to total volume.** For example, a mass of 75 grams, (g), measured using a top-loading chemical balance, may have a volume of 15 cubic centimetres, (cm^3). This mass has a density of 75 g/15 cm^3 or 5.0 g/cm^3. Each cubic centimetre has a mass of 5.0 grams.

$$\text{density} = \text{mass/volume, or } d = m/V \tag{1.1}$$

If you know the density of a substance, you know its mass per unit volume. You can predict the mass of a structure made of this material, once you know the volume of material in the structure. The total mass of an object is the mass of unit volume times the number of unit volumes it contains, i.e.

$$\text{total mass} = \text{density} \times \text{total volume or } m = d \times V \tag{1.2}$$

A typical experiment to compare densities needs a **chemical balance** and a transparent **measuring cylinder** marked with a volume scale. We find the volume of a solid by finding the volume of liquid it displaces (pushes aside), when totally immersed. The readings might be:

Mass of a lump of material X	= 45 g
Volume of liquid in the cylinder	= 67 cm^3
Volume of liquid and lump X	= 77 cm^3
Volume of lump $X = (77 - 67)$ cm^3	= 10 cm^3
Density of X = mass/volume = 45 g/10 cm^3 = 4.5 g/cm^3	

Mass of lump of material Y	= 63 g
Volume of liquid in the cylinder	= 48 cm^3
Volume of liquid and lump Y	= 66 cm^3
Volume of lump $Y = (66 - 48)$ cm^3	= 18 cm^3
Density of Y = mass/volume = 63 g/18 cm^3 = 3.5 g/cm^3	

This shows that X is denser than Y in the ratio 4.5/3.5, or 9/7. Table 1.1 shows the densities of several substances. The well-known density of water is 1000 kg/m^3. This is also 1 g/cm^3 as shown in the next column. Notice that all the values in the left column are 1000 times bigger than in the right column. This is entirely due to the difference of unit.

The **density of a solid remains the same** even in space or on a distant planet. If neither the mass of an iron bar nor its volume changes in space, its density will remain the same.

Self test

1 m^3 contains 1 000 000 cm^3. Material Y has a density of 3.5 g/cm^3. How many grams of Y fit into 1 cubic metre? Show that this amounts to 3500 kg. (Remember kilo– means $\times$ 1000; so 1 kg is 1000 g).

Table 1.1 Some densities of common substances

Units	kg/m³	g/cm³
Solids		
Lead	11 800	11.8
Copper	8900	8.9
Aluminium	2700	2.7
Ice	920	0.92
Wood	600–1100	0.6–1.1
Cork	240	0.24
Liquids		
Mercury	13 600	13.6
Water	1000	1.0
Petrol	680–720	0.68–0.72
Gases		
Air at 0°C at sea level	1.29	1.29×10^{-3}
Hydrogen at 0°C at sea level	9×10^{-2}	9×10^{-5}

1.4 Experimental results

Putting numbers to the observations we make in science enables us to record conditions more precisely. For the numbers to be accurate, the apparatus often needs careful adjustment and the readings must be repeatable. Repeated readings are sometimes made to see how much variation there is in them, and to obtain a reliable average for each reading.

1.5 Mass and weight

The invisible force of gravity on an object, measured with a **spring balance**, depends on its mass. If the mass is doubled, the force of gravity also doubles. **Weight is the force on a mass due to gravity**, and on Earth the force is about 10 N for every kilogram of mass. For example, you might weigh yourself on bathroom scales and find you weigh 450 N. As each kilogram on Earth weighs about ten newtons, your mass would be about 45 kg.

There is more about weight and mass in section 7.2. The property relating them is:

$$\text{Weight (N)} = \text{mass (kg)} \times \text{force of gravity on 1 kg, or } W = m \times g \qquad (1.3)$$

Sometimes the weight of an object is represented by a single downward arrow where this invisible force appears to act. There is a point on each object, known as its **centre of gravity**, (C of G), or **centre of mass**, through which the weight of the whole object is considered to act.

1.6 The concept of pressure

Earlier we suggested that a thumb tack or drawing pin would sink into plasticene more easily than your finger. How do we explain that?

Figure 1.1 Design of a thumb tack

Look at the drawing pin or thumb tack in Fig. 1.1. The tip is sharp and pointed to make it sink into a bulletin board, but the head is flat and spread out so that it does not sink far into your thumb. **The same force acts at each end,** but where it is **concentrated on a smaller area, the pressure is greater** and the pin sinks in. So what is pressure?

Solids, liquids and gases all exert pressure on whatever supports them. Masses weigh down on the Earth's surface, and we measure their pressure by the size of their weight acting down on each unit of surface area.

Pressure is a force acting over an area.

$$\text{Pressure} = \frac{\text{force (N)}}{\text{area (m}^2)} \text{ or } P = \frac{F}{A} \tag{1.4}$$

> **Example**
> A patio paving slab of dimensions 0.5 m × 0.5 m × 0.05 m weighs 625 N. Its weight is evenly supported on an area of 0.5 m × 0.5 m on the ground (See Fig. 1.2a). The pressure exerted by the slab = force/area, so
>
> $$\text{Pressure } P = \frac{F}{A} = \frac{625 \text{ N}}{0.5 \text{ m} \times 0.5 \text{ m}} = 2500 \text{ N/m}^2.$$

One newton per square metre is called one pascal (Pa). This is the SI unit of pressure. There are other units too, so it is important to put down the unit when recording a pressure. The number alone would be meaningless.

In how many ways could the pressure beneath the slab have been increased? To raise the pressure we would need more force per unit area. We could make the force bigger, or the area smaller.

> When the patio slab rests on one of its smaller sides of area 0.5 m × 0.05 m, as in Fig. 1.2b, the pressure is:
>
> $$P = \frac{F}{A} = \frac{625 \text{ N}}{0.5 \text{ m} \times 0.05 \text{ m}} = 25\,000 \text{ N/m}^2 = 25 \text{ kPa.}$$

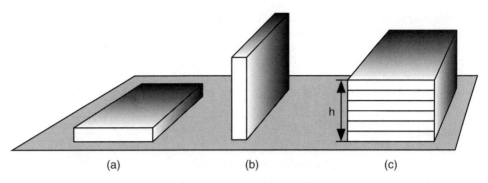

Figure 1.2 Rectangular paving slabs

The pressure becomes ten times bigger when the supporting surface area is only one tenth as big. This explains why a sharp engraving or cutting tool is so effective. The applied force may not be large, but the area on which it is applied is very small, so the pressure on the surface is very great and it sinks in, just like the tip of a thumb tack.

1.7 Pressure below solids and liquids

A stack of three paving slabs will be three times as heavy as one, and the force on its base will be trebled. The pressure will also increase threefold, as it has the same base area as before. A taller stack will have more slabs, and exert still more pressure (see Fig. 1.2c). In general, **the pressure will be in direct proportion to the height, h, of the stack.**

A second important factor or 'variable' is the density of the material in the slab. What would happen to the weight of a stack of patio slabs if their densities were all doubled? How would this affect the pressure beneath the stack? Both the weight and the pressure would double. The **pressure exerted by a stack is in direct proportion to the density, ρ, of its material.**

Another variable affecting the pressure is the strength of the Earth's gravity, without which the slabs would not weigh down on the Earth's surface at all. Doubling the Earth's gravitational pull would double the weight of each patio slab, and therefore double the pressure beneath it. **The pressure is directly proportional to the strength, g, of the Earth's gravity,** which is about ten newtons of force per kilogram of matter. So there are three independent variables which determine the pressure, namely the height of the stack, h, its density, ρ, and the strength, g, of the Earth's gravity. The pressure is proportional to all three variables. So in Fig. 1.2c

The pressure beneath the stack is its weight/base area.	
The volume = base area × height	$= A \times h$
The mass = volume × density	$= A \times h \times \rho$
The weight = mass × force of gravity on unit mass	$= A \times h \times \rho \times g$
The pressure $= \dfrac{\text{force}}{\text{area}} = \dfrac{A \times h \times \rho \times g}{A}$	$= h \times \rho \times g$

Pressure = height × density × gravitational field strength or $h \times \rho \times g$ (1.5)

The result applies equally to vertical columns of both solids and liquids, which are virtually incompressible so that their densities are almost constant. It is important for building surveyors and deep sea divers alike. We might have expected the area of a swimming pool or building to affect the pressure at its base; but the area factor, A, cancels out of the pressure calculation, and the pressure is independent of the area. A narrow vertical tube of liquid can exert the same pressure as a reservoir of equal depth.

Example
How much does the pressure increase as a diver descends 30 m in water?
(Density of water $= 1000$ kg/m^3, and g $= 10$ N/kg).
Extra pressure $\quad = h \times \rho \times g = 30$ m $\times 1000$ kg/m$^3 \times 10$ N/kg
$\quad\quad\quad\quad\quad\quad = 300\,000$ Pa, or 300 kPa, (i.e. about $3 \times$ atmospheric pressure).

1.8 More about liquids

A diver carrying a handheld pressure gauge can confirm that its reading increases uniformly with depth below the surface of water. However, at a given depth, the meter can be turned on its side or even upside down without any change in the reading. **The pressure at a given depth in a liquid is exerted equally in all directions,** including upwards as well as downwards and sideways.

In Fig. 1.3, a hollow cylinder of water has three similar holes at different depths. At each hole, the water escapes from the higher pressure inside the cylinder to the lower pressure outside. The pressure difference pushes water from each hole. Can you explain at which hole the water escapes fastest? There is more about hydraulic pressures in chapter 9.

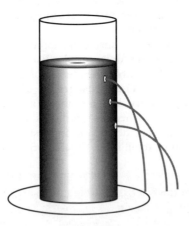

Figure 1.3

1.9 Fluids

The word 'fluid' applies to liquids and gases, both of which exert pressure at a point equally in all directions. As we have seen above, gases can be compressed, and in normal conditions they have much smaller densities than liquids. When they are mixed up together, the less dense gases rise to the top above the denser liquids. But some gases, like oxygen for instance, can dissolve in liquids like water in limited quantities. Fish need this dissolved oxygen in order to survive.

In all fluids, the pressure increases with depth.

1.10 Pressure in the Earth's atmosphere

The atmosphere is denser near the Earth's surface, and it exerts more pressure than at altitude. The main reason for this is that there is simply more gas weighing down from above near sea level than there is at a point higher up. The cabin of a passenger aircraft is pressurised as the aircraft gains altitude, because the air outside is too 'thin' for comfort.

Atmospheric pressure is measured with a barometer. Some barometers enclose a mercury column which exerts the same pressure at its base as the atmosphere; others detect the movement of the thin wall of a sealed metal box as the atmospheric pressure changes. These, having no liquids, are called aneroid barometers.

A sealed metal can collapses when the air is pumped out of it, because it is not strong enough to withstand the external atmospheric pressure. When the air inside the can exerts an equal pressure from within, the walls do not collapse. Similarly, the air inside a house exerts the same pressure as the atmosphere outside. Without this balance of pressure, the house might collapse.

A number of gases contribute to the pressure of air in our atmosphere. About 80 per cent of air is nitrogen, and most of the rest is oxygen. But other gases are also important. Carbon dioxide, for example, is essential for photosynthesis in plants. There is more about some other gases in later chapters.

Water vapour in the atmosphere is due to the evaporation of water from the ocean. It gives rise to humidity, and exerts a small pressure. As it rises with thermal upcurrents, the cooling vapour condenses into tiny water droplets, which become visible as clouds. Only when many droplets collect together to make larger, heavier, raindrops do we experience rain, which completes the cycle when it flows back into the ocean.

Questions

1 (a) What is the pressure when a force of 10 N acts on an area of 2 square metres?
   (b) If the atmospheric pressure is 100 000 Pa, what force does it exert on a plate glass shop window of dimensions 3 m × 2 m?
   (c) What prevents the window in part (b) from being pushed in?
2 A solid rectangular block 4.0 m × 2.0 m × 1.0 m in size, weighs 50 kN. One face rests on the ground.
   (a) What is the greatest pressure it can exert?
   (b) What is the least pressure it can exert on the ground?
3 It is said that a ballet dancer weighing 500 N standing on one toe can exert more pressure than an elephant weighing 60 kN standing on its four feet. Do some calculations to see if you agree or not. Explain your reasoning.
4 Explain each of the following:
   (a) Liquid does not pour easily from a sealed tin pierced by one small hole.
   (b) The pressure in the atmosphere decreases as altitude increases.

What have you learnt?

1 Can you explain that the densities of a copper cylinder and a copper cube are the same?
2 Can you write a sentence to explain the difference between force and pressure?
3 Can you calculate a pressure, given the force which acts and the area it acts upon?
4 Do you know that it is difficult to change a liquid's density by compressing it?
5 Can you explain why wide tyres are fitted to tractors?
6 Do you know where the pressure of the atmosphere surrounding the Earth is greatest?
7 Can you name the apparatus used to measure the atmospheric pressure?
8 Do you know how to calculate the pressure at the bottom of a vertical column of a liquid or solid?

② Strength of solids

Objectives

When you have completed this chapter you should
- be able to use the words in **bold** accurately
- be able to answer problems involving Hooke's law
- have a picture of the layers of molecules in a crystal
- understand what changes occur when the crystal is stretched or deformed
- understand graphs of extending force plotted against extension for elastic objects.

2.1 Introduction

If you are a climber, you need a **strong** rope. That means it will withstand a large force without breaking. A good rope is not too heavy, even in freezing or rainy conditions. It must be **tough**, as a **brittle** rope **may snap** when hit by a rock; and it must not fray when rubbed against a hard surface. **'Tough' means 'not brittle'**, and a tough material often extends and narrows at a weak point before breaking.

The photograph in Fig. 2.1 shows mountain rescue training. Here the climbers rely on the strength of ropes to prevent a misfortune turning into a crisis. For safety reasons a rope must be able to withstand a force several times greater than the weight of the object (climber or stretcher) it supports. The ropes have both **elasticity** and **strength**. Their elasticity enables them to stretch as they bring a falling climber to rest gradually reducing the risk of injury.

In choosing a rope, a climber knows that a thin rope will be lighter than a thicker one of the same material, and it will also stretch more easily. But a thinner rope may not be strong enough and may break if over-loaded. The **strength** of a rope is defined as the **least force needed to break it**.

2.2 Elastic materials

The elasticity or **stiffness** of a rope, k, is the amount of force needed to extend the rope by one unit of length. So a sample which is easy to stretch has a small k value, and a stiffer sample has a larger k value. Force is measured in newtons, N, length in metres, m, and k is usually measured in newtons per metre, (N/m), or in N/cm.

Figure 2.1 Survival may depend on safety procedures and the strength of the climbers' ropes. (Nick Parks)

Provided the force is not too great, the increase in length or **extension (see Table 2.1) of an elastic object increases uniformly with the applied stretching force.** Their ratio may be written:

force/extension = stiffness (k), or $F/x = k$, so that $F = k \cdot x$ (2.1)

This property of elastic materials is called **Hooke's law,** and it means that up to a certain limit, the extension of a sample will double when the force stretching it is doubled. The value of k is smaller for a long rope, which is easier to stretch than a shorter one.

2.3 Experimental tests

To test the behaviour of metal springs, a steel spring may be extended with known forces, **using a spring balance** as in Fig. 2.2. As the force is increased, the spring gets longer. Each force is noted, and the length of the spring is measured with a ruler. The readings are listed in two columns in Table 2.1. A third column is used in the table of values to show the extension of the spring, which is its total length minus its original (unloaded) length.

Fig. 2.3 shows the force and extension results for the steel spring plotted as a graph. The graph is a straight line through the origin (0,0). It confirms that Hooke's law applies to the spring, so the extension is in direct proportion to the force. The stiffness, k, or **'spring constant'** is shown in the last column of Table 2.1. This shows the ratio of force to extension at any point on the graph is 2 N/cm, or 200 N/m. The constant can also be found from the **gradient of the graph**. The gradient is the increase in the y direction divided by the corresponding increase in the x direction (see Fig. 2.3).

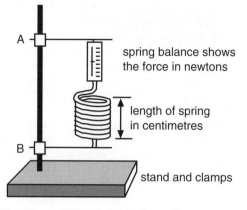

Figure 2.2 The distance AB is varied, and the readings are recorded in Table 2.1

Table 2.1 Readings taken with the steel spring shown in Fig. 2.2.

Force (N)	Length of spring (cm)	Extension (cm)	Force/extension
0	10	0	k (N/cm)
2	11	11 − 10 = 1	2/1 = 2
4	12	12 − 10 = 2	4/2 = 2
6	13	13 − 10 = 3	6/3 = 2
8	14	14 − 10 = 4	8/4 = 2

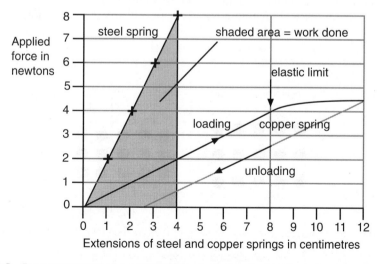

Figure 2.3 Force against extension graphs are shown for two metals. The spring constant, k, for steel is the gradient of the graph. For the steel spring $k = 8$ N/4 cm. So $k = 2$ N/cm, or 200 N/m

A copper spring the same size as the steel one can be made by winding the same gauge copper wire onto a rod of the right diameter. With the springs having an equal number of turns, a loop is made at each end, and the experiment is carried out as before. When the results are plotted, the graph is again a straight line for the smaller forces. But, this time, it curves towards the extension axis after reaching a point called the '**elastic limit**'. After being extended beyond its elastic limit, the copper spring stops behaving elastically, and it does not fully return to its original length when the force is removed (see Fig. 2.3).

The graph for the steel spring is steeper than for the copper spring. This means that the steel spring has the greater stiffness, and the copper spring is the easier to stretch. Notice that the steel spring was not extended as far as its elastic limit, beyond which it too would have been permanently extended. You can see the importance of designing machine parts so that they are not stressed to their elastic limits, and why steel parts are often preferred to copper or aluminium. Equation 2.1 applies generally to solids before their elastic limit is reached.

2.4 Molecular forces in solids

The springs experiment shows that different substances of the same shape have different stiffness. This is also found from tests on long straight wires of different substances. One of the reasons for different stiffnesses concerns the tiny forces between molecules inside the materials. X-ray tests show that metals are made of minute crystals. In each crystal the molecules are arranged in regular layers which are spaced on top of one another with regular spacing (see Fig. 2.4a). To stretch the material elastically, the layers may be pulled a bit wider apart; and to compress it, the layers may be moved closer together. But when the molecules are pushed closer together they repel each other to move apart again; and when they are pulled apart, they attract each other. Another way of stretching a crystal elastically is shown in Fig. 2.5b, where the layers of molecules are displaced parallel to each other. But this too results in restoring forces trying to return the molecules to their original positions. The intermolecular forces vary with the substance, and affect both its strength and stiffness.

When a small force stretching a wire is removed, the layers of molecules in the

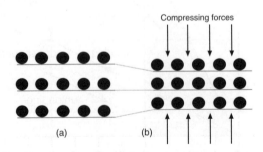

Figure 2.4 Molecules are too small to be seen. They are represented here by large circles to show that:
(a) layers of molecules in a crystal are evenly spaced.
(b) as the crystal is compressed the layers move closer.
The figure is enlarged about 100 million times

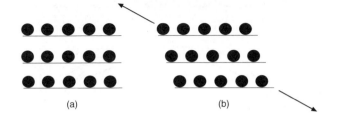

Figure 2.5 Layers of molecules are represented (a) in an unstretched crystal, and (b) when displaced in their own plane by stretching forces

wire return to their original positions. But if the force extends the wire **beyond its elastic limit**, some molecular layers slide over others, and when the force is removed, the displaced layers do not slide back to their original positions. In this way the wire is permanently stretched.

2.5 Energy stored in an elastic material

Some energy is needed to stretch an elastic material. If its elastic limit is not reached, the substance returns the energy when it returns to its original shape. But beyond the elastic limit permanent deformation occurs, and some of the energy put into the material is not returned. It could, for example, become heat in the deformed material.

Energy is the capacity for doing work, and it has the same units as work (joules). To find the work done by a force as it pulls or pushes something along, we calculate the product:

Work done = force × distance moved, or $W = F . d$ $\qquad$ (2.2)

> **Example**
> A horse on a towpath hauls a canal barge along the canal at a steady rate of 2 km/h. The force it exerts on the barge is 400 N. How much work is done on the barge every hour?
>
> **Answer**
> In one hour the barge moves 2000 m. The work done in one hour is W, where:
>
> W = force × distance moved = 400 N × 2000 m = 800 000 J.

In the graph of force against extension, the work done as the force increases is found from the area between the graph and the extension axis. For an elastic material, the area under the graph is triangular, and the **work done = average force × final extension**.

= (1/2) final force × final extension

Another way of writing this is: work $\quad$ = (1/2) $(k . x) x = (1/2) k . x^2$ $\qquad$ (2.3)

So the energy needed to stretch the steel spring in Fig. 2.3, is (1/2) 8.0 N × 0.04 m = 0.16 J.

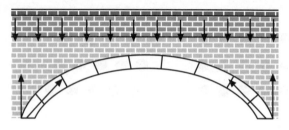

Figure 2.6 This figure shows the forces from the ground supporting a stone arch, and the vertical load forces pushing it down. The forces combine to compress the stones and hold the structure together

2.6 Strengths of structures

The age of some of the oldest river bridges points to the strength of stone arches. The stones are held in compression, as the forces acting on the bridge push them together (see Fig. 2.6). On the other hand, steel is strong both in compression and in tension. Suspension bridges use these properties along with the stiffness of steel components (see Fig. 2.7). Bridges can carry larger loads if the components they are made from are stressed so as to use their greatest strengths.

In Fig. 2.7a all the components supporting the roadway are being stretched, except for the two thick vertical towers. The towers are being compressed. Fig. 2.7b shows the forces acting at the top of one of the towers in Fig. 2.7a. The cable is in tension, pulling downwards and sideways. But the sideways effects to the left and right cancel out. The tower pushes upwards on this junction enough to balance the two downward forces of the cable. So all the forces acting on the top of the tower cancel each other out. This is also true at any other junction between cables in the bridge. In a stable structure the forces at each point must balance, and so must all the external forces acting on the structure.

___ **Self test** ___

Explain the forces acting at a junction of the structure in Fig. 2.8.

Some of the principles of bridge design and construction can be explored in the laboratory using materials like paper, string, and drinking straws. It is best to discover whether each material is stronger in compression or in tension, and then to design a structure which uses each material to best advantage. Try organising a competition in

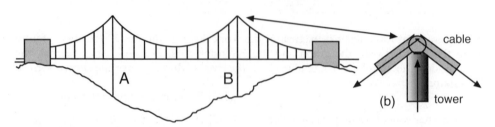

Figure 2.7 This figure shows (a) a suspension bridge supported by towers A and B and (b) the forces at the top of tower B where the downward pull of the cable on each side is balanced by the upward push of the tower

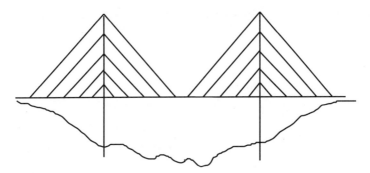

Figure 2.8 *This figure shows another bridge design. (see also Fig 5.5)*

the laboratory between teams. Specify the gap to be spanned and the materials available. The winners will produce the strongest bridge with the least material.

2.7 Some uses of strong elastic materials

A moving mass has **kinetic energy**, or energy of motion. Sometimes it is necessary to transform this energy safely as the mass is brought to rest. A Formula 1 racing car may go out of control at 200 km/h, and gravel traps beside the race track may convert some of the car's kinetic energy into work done on the gravel. The remaining energy is converted into work done in compressing and distorting a barrier of car tyres, and in bending the car.

The picture in Fig. 2.9 shows a strong elastic material used in nature. Best seen on a dewy sunlit autumn morning, the spider's web is used to trap flying insects. Its central area presents a barrier to their flight paths, and it is often supported by surprisingly few diagonal ties to nearby stems or leaves of plants. The spider's thread is very fine,

Figure 2.9 *The spider's web is made of a very fine, strong material*

yet the web can stretch and absorb the energy of a flying insect without breaking. If this material were as thick as a climber's rope, we would see that it is very strong indeed.

The spider's craft
How do you think a spider constructs a suspension bridge from an apple tree to a shrub more than four metres away over running water and at night?

The technique it uses is remarkable. First the spider lets out a very fine line, which drifts in the breeze. Eventually the line snags an obstacle at the far end and sticks to it. Then the spider attaches a thicker line at the start and moves along the suspended line taking it in, and creating as it goes the new thicker line from start to finish. These are the first steps in the creation of a structural masterpiece.

In a shipping emergency a large vessel is drawn by a tug using a towline in heavy seas. Although the towline is a thick steel cable, there is a serious risk of it being stretched beyond its elastic limit as the vessels move in the ocean swell. If the cable should break, there would be a risk to those on the decks of both ships as the free ends flew back. The risk of the line breaking can be reduced if a longer towline is used. The stretching brought about by the pitching and rolling of the ships can then be shared out along many metres of cable so that its elastic limit is not reached.

Questions

1 Strong, tough, brittle, stiff; which of these descriptions apply to:
   (a) a biscuit, (b) an elastic band, (c) window glass, and (d) a fibre glass canoe?
2 Use Fig. 2.3 to find the spring constant, k, of the copper spring.
3 Use Fig. 2.3 to find the energy stored in the copper spring when it was extended to its elastic limit but not beyond it.
4 The force acting on a spring balance is shown by a pointer on a scale of uniformly-spaced numbers. Whose law applies to the extensions of the spring in this device?
5 (a) Draw a diagram to show how molecules could be arranged in a crystal.
   (b) How can molecules in a crystal move as it is stretched elastically?
6 Tough, brittle, stiff, weak, strong; which qualities are needed in a cable used to lower a man from an air-sea rescue helicopter?

What have you learnt?

1 Can you name materials which are (a) tough, (b) brittle, (c) strong, and (d) stiff?
2 Can you explain Hooke's law?
3 Do you know what is meant by the elastic limit?
4 Can you describe the graph of force against extension for an elastic spring?
5 Can you find the work done stretching a spring?
6 Can you explain the ways forces balance at any point in a stable structure?
7 Do you understand that a crystal has molecules arranged in regular layers?

3 Particles in motion

3.1 Introduction

Liquid water can be cooled down so that it freezes and turns into ice. Or it can be heated up until it boils and turns into steam. There are three different states of the same material, ice, water and steam. For substances in general these three states are called:

(a) **solid**, (b) **liquid**, and (c) **vapour** or **gas**.

3.2 The three states of matter

Fig. 3.1 shows the differences between the three states of matter. A solid has a fixed shape which is not affected by the shape of its container. A liquid takes the shape of the container, up to the level of the liquid's surface. A gas fills the entire container.

The volume of liquid water is about the same as the volume of ice which can be formed from it. But when it boils and becomes steam in a kettle, its volume is nearly 2000 times greater. When any liquid is turned into a gas, there is always a considerable increase in its volume.

The smallest particles of any substance are called **atoms**, or sometimes **molecules**, which are groups of atoms. Water is made of molecules, each containing two atoms of hydrogen and one atom of oxygen. Its chemical symbol is H_2O.

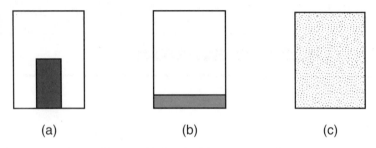

(a)　　　　　　　　(b)　　　　　　　　(c)

Figure 3.1 The containers hold (a) a solid, (b) a liquid and (c) a gas. The molecules in the solid vibrate about their fixed positions. In (b) they are free to move inside the liquid. In the gas they move faster and freely anywhere in their container, colliding with its walls and with each other

In the solid state, each of the water molecules is in a fixed position relative to those around it. This is why a piece of ice has a fixed shape. Forces hold the molecules in place in the solid, which can only be reshaped with great difficulty. So, in Fig. 3.1a, the molecules can vibrate while keeping their regular crystalline arrangement.

When ice has **melted** to form a liquid, its solid shape is lost, and the water molecules move about more freely. They have more energy than they had in the solid. So energy must have been supplied to the ice to enable it to change into liquid water as in Fig. 3.1b.

When the water has boiled and vaporized, the molecules have more energy still. This enables their distances from each other to increase considerably. They are in rapid, random motion, and their movement is restricted only by the walls of their container (see Fig. 3.1c). The extra space between the molecules of a gas means that a gas can be compressed, whereas the small space between molecules in a solid or liquid makes it very difficult to compress any substance in these states.

3.3 The kinetic theory of matter

This is the theory that molecules in a liquid or gas are in rapid, random motion. The molecules are imagined to move freely, like miniature snooker balls. From time to time they bump into each other or the walls of their container. In a mixture of gases, all the molecules are imagined to have, on average, the same amount of kinetic energy. What this means is that the ones with less mass have greater average speeds. So in a mixture, hydrogen molecules would on average move faster than oxygen molecules, which would on average move faster than the more massive carbon dioxide molecules. If the temperature of the mixture were raised, the average molecular energies and speeds would increase.

3.4 Diffusion

When you cut up an onion the smell is not immediately detectable a few metres away. But soon the molecules which carry the smell will spread throughout the kitchen even in still air conditions. The delay, between cutting the onion and smelling it some metres away is caused by air molecules which collide with molecules emerging from the onion. The resulting chaotic mixing as one gas or vapour spreads through others is called diffusion. It can also be demonstrated in liquids when a coloured liquid diffuses into a clear one over a period of time.

The kinetic theory enables various properties of both liquids and gases to be understood. The next section gives some examples.

3.5 Evaporation and boiling

In a liquid at a particular temperature the average energy of each molecule over a long period is the same. At any instant, however, there will be a spread of molecular speeds; some will be moving faster than others. A molecule which bumps into the wall of the container will bounce back into the body of the liquid. One which moves up to the liquid surface may break out before being pulled back into the liquid by the forces of intermolecular attraction. But if the molecule is travelling particularly fast (well above the average speed), it might manage to break free from the liquid altogether. This is the process of **evaporation**. It occurs only at the liquid's surface.

Now think of what happens if the liquid temperature is raised. The average speed of the molecules increases with their kinetic energy, so that more molecules have enough speed to escape from the liquid surface; the rate of evaporation increases. Eventually, at a high enough temperature, so many molecules escape that they sweep away the air molecules above the surface. It is even possible for a group of molecules to form a bubble of vapour in the body of the liquid. This process, called **boiling**, occurs throughout the liquid at one temperature called the **boiling point**, when the conversion from liquid to vapour is rapid.

3.6 Boyle's law for gases

The apparatus illustrated in Fig. 3.2 enables the pressure on a sample of air to be varied by means of a pump, so as to change its volume. Corresponding readings of the pressure and volume are taken at the constant temperature of the laboratory. It is found that the pressure × volume remains the same for each pair of readings. A graph of the pressure plotted against 1/volume in Fig. 3.3 is a straight line through the origin, showing that the pressure is **inversely proportional** to the volume. The results agree with **Boyle's Law** which says that, **for a fixed mass of gas at constant temperature, the product:**

Pressure × Volume is a constant or, $P \times V =$ constant   (3.1)

3.7 Kinetic theory and Boyle's law

The pressure exerted by a gas on the walls of its container is thought to be due to the collisions between the gas molecules and the walls. If the gas temperature remains fixed, the average speed of the molecules is fixed.

If the volume of a fixed amount of gas is halved, there will be twice as many molecules in a given volume, so there will be twice as many collisions per second on each square centimetre of wall. The pressure will be doubled. With the volume halved and

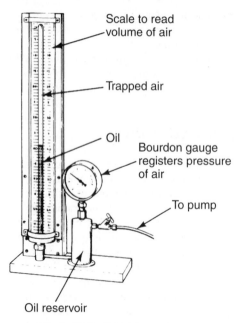

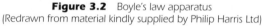

Figure 3.2 Boyle's law apparatus
(Redrawn from material kindly supplied by Philip Harris Ltd)

the pressure doubled, the product pressure × volume will remain unchanged, as long as the temperature is the same.

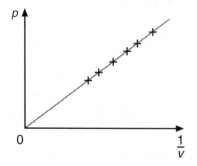

Figure 3.3 A graph of pressure plotted against 1/volume for a gas using results obtained with the apparatus of Fig. 3.2

3.8 The gas pressure changes with temperature

Fig. 3.4 illustrates the apparatus used to find how the pressure in a gas changes with temperature, while its volume remains constant. The pressure is indicated by the Bourdon gauge and the temperature is shown by the thermometer in water surrounding the flask. The volume of trapped air in the apparatus is virtually constant.

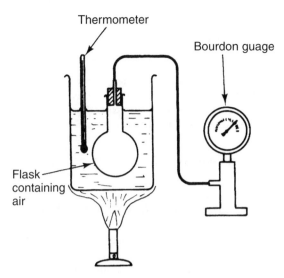

Thermometer

Bourdon guage

Flask
containing
air

Figure 3.4 *An apparatus for measuring how the pressure of a fixed volume of gas changes as the temperature changes*

The results indicated by the graph in Fig. 3.5 show that the pressure falls as the temperature falls, and that the pressure would eventually reach zero if the gas did not first liquefy. The zero of pressure, when molecules would cease to move, would be reached for all gases at about −273°C. We cannot imagine a gas exerting less than zero pressure, so we think of −273°C as the lowest possible temperature. It is called the **absolute zero**, or zero degrees kelvin, and written 0 K.

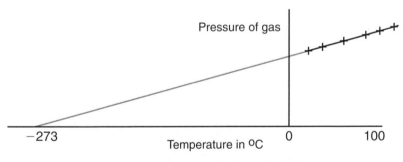

Pressure of gas

−273

0

100

Temperature in °C

Figure 3.5 *A graph of results obtained with the apparatus of Fig. 3.4*

3.9 Using the kelvin temperature scale

Temperature intervals on this kelvin scale are chosen so that a rise of 10 K is the same as a rise of 10°C. So 0°C, which is 273 degrees above the absolute zero, is 273 K. To convert a temperature from degrees celsius to its kelvin equivalent, you just add 273. For example 50°C = (50 + 273) K = 323 K. To convert a kelvin temperature to its celsius equivalent, you subtract 273. For example 400 K = (400 − 273)°C = 127°C.

Notice that in Fig. 3.5 the graph passes through the origin if the temperature axis is marked in kelvin temperatures. This is true for all gases. So **for a fixed mass of gas at constant volume, the pressure is proportional to the absolute temperature**. This is known as the **pressure law for gas**es. If P is the pressure and T the absolute (kelvin) temperature, then

$$P/T = \text{constant}, A, \text{ at constant volume, or } P = A \times T \tag{3.2}$$

3.10 The gas volume changes with temperature

The apparatus illustrated in Fig. 3.6 may be used to find out how the volume of a gas changes with temperature at constant pressure. The air trapped in the tube by the index of concentrated sulphuric acid remains free of water vapour, and it is also under atmospheric pressure throughout the experiment.

The volume of the gas is indicated by the length of the air column, and the variation of volume with temperature is indicated by Fig. 3.7. This shows that the volume decreases as the temperature decreases, and the volume would become zero at −273°C if the gas did not first liquefy. So in the same way as before we may say that, **for a fixed mass of gas at constant pressure, the volume is proportional to its absolute (kelvin) temperature**. In symbols, if V is the volume of gas at T K, $V/T = \text{constant}, B,$ or

$$V = B \times T \tag{3.3}$$

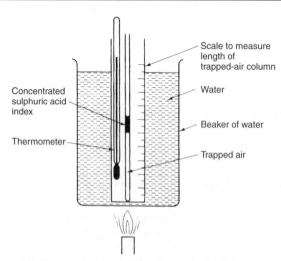

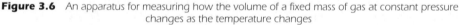

Figure 3.6 An apparatus for measuring how the volume of a fixed mass of gas at constant pressure changes as the temperature changes

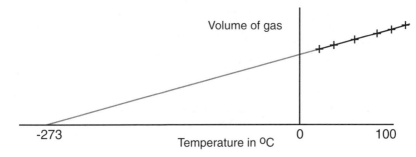

Figure 3.7 *A graph of results obtained with the apparatus of Fig. 3.6*

3.11 The ideal gas equation

Each of the three gas equations (3.1), (3.2), and (3.3) apply to a fixed mass of gas in particular conditions – for example, constant temperature. In any change of conditions the equations may be combined into one equation which states that:

for a fixed mass of gas

$$\frac{P_1 \times V_1}{T_1} = \frac{P_2 \times V_2}{T_2} \text{ where } T \text{ is measured in kelvins} \tag{3.4}$$

This is called the ideal gas equation. The starting conditions are labelled with subscript 1, and the final conditions with subscript 2.

Example
If one litre of oxygen at 22°C and one atmosphere pressure is heated to 98°C in a container at 0.8 atmospheres pressure, what will be its final volume?

Answer
Use equation (3.4). Then

$$\frac{1 \text{ at} \times 1 \text{ li}}{(273 + 22) \text{ K}} = \frac{0.8 \text{ at} \times V}{(273 + 98)} \text{ K}.$$

So the final volume

$$V = \frac{1 \times 371}{0.8 \times 295} = 1.6 \text{ litres.}$$

Notice we have used atmospheres of pressure on each side, and litres of volume, but T must be in kelvins (K).

Self test
Find the final pressure of air if 2 litres at one atmosphere pressure and 27°C are expanded to 3 litres at 87°C.

Answer

Use equation 3.4. Then

$$\frac{1 \text{ at} \times 2 \text{ li}}{300 \text{ K}} = \frac{P_2 \times 3 \text{ li}}{360 \text{ K}}.$$

From this, $1/150 = P_2/120$, and the final pressure is $P_2 = 120/150 = 0.8$ atmospheres.

3.12 The pressure cooker

In an earlier section boiling was said to occur if enough high speed molecules escaped from the liquid to sweep away any air molecules above the surface. If there are more air molecules present because the air pressure is greater, then a higher temperature is needed for boiling to occur.

At a pressure of one atmosphere, water boils at 100°C, and at two atmospheres it boils at about 120°C. At this higher temperature the cooking times for food are considerably reduced. Cooking involves physical and chemical changes which happen more rapidly at higher temperatures.

Inside a pressure cooker the pressure is allowed to reach about two atmospheres, and at this pressure cooking is much more rapid than at normal atmospheric pressure. However, mountaineers find that cooking is slower at high altitude where water boils at lower temperatures. At the top of Mount Everest, water would boil at only 70°C.

3.13 Refrigerators

When you pump up a bicycle tyre, the end of the pump nearest the tyre gets hot. The reason is that moving gas molecules in the pump collide with the moving piston during compression, and bounce off faster. So the gas warms up. But if it is decompressed, the gas cools down. So reducing the pressure on a gas, and enabling it to expand, can be used for refrigeration.

Another change which results in cooling is the evaporation of a liquid. When the faster molecules break through the liquid surface into the space above, the average energy of the molecules which remain in the liquid will be less. So the liquid becomes cooler. A domestic refrigerator decompresses and evaporates a liquid in a sealed unit, so as to keep food cool in the space nearby.

Questions

1. In which state of matter do molecules have least energy?
2. In which state of matter is a substance most easily compressed?
3. In all gases at a given temperature what averages out to be the same for each molecule?
4. Explain one change which occurs when a liquid or gas is heated.
5. What happens to the rate of evaporation of a liquid as it is heated?
6. Explain why an evaporating liquid becomes cooler if no energy is supplied to it.

7   A pump has compressed some air into one third of its original volume.
   (a)   Why is the gas warmer immediately after the compression?
   (b)   When the gas has cooled to its original temperature, how will its pressure compare with the original pressure?
   (c)   Whose law did you apply in your answer to part (b)?
8   Air at 27°C is heated in a fixed volume until its temperature becomes 77°C.
   (a)   Convert each of these temperatures into their kelvin equivalents.
   (b)   By what factor will the pressure of the air change as a result of the rise in temperature?

What have you learnt?

1   Do you know the three states of matter?
2   Do you know how the motion of molecules can explain the properties of fluids?
3   Do you know what Boyle's law says about gases?
4   Can you convert 100°C into its equivalent kelvin temperature?
5   Do you know how changes of temperature can affect the pressure or volume of a gas?
6   Can you explain the differences between evaporation and boiling?
7   Can you explain the principles of (a) the pressure cooker and (b) the refrigerator?

4 Motion

Objectives

After this chapter you should
- understand what is meant by average speed
- understand travel graphs of displacements against time
- know the difference between speed and velocity
- know how to measure a velocity
- be able to use graphs of velocity against time to find accelerations and distances gone
- know what acceleration means
- know how to measure accelerations
- know that a mass in uniform circular motion accelerates.

4.1 Introduction

Today we travelled 300 miles in the car in 12 hours. So our average speed must have been 300/12 or 25 miles per hour. To find the **average speed** you work out

$$\text{average speed} = \frac{\text{total distance gone}}{\text{total time taken}} \qquad (4.1)$$

It does not sound as if we moved very fast, but we travelled at 70 mph on the motorway, except for the bits where there were road works. Of course, the times spent having lunch with my aunt and her family are included in the twelve hours. So a good deal of time was spent travelling faster than 25 mph, but for much of the time the car was standing still.

Speed is an example of a **scalar quantity**. Scalars have **size but no particular direction**. Mass, temperature and energy are all examples of scalars.

A velocity is a speed in a certain direction. Speed is the size of the velocity whose direction can be shown on a map by an arrow, or written as a compass bearing e.g. 25 m/s in the direction due West. Quantities having both **size and direction** are **vector quantities**, so velocity is a vector quantity, as are force and momentum.

4.2 Displacements

A tape measure laid along a straight line contains a scale of uniformly spaced numbers showing the **displacement** of any position from the end. A displacement can be measured in metres. A long jumper leaps up and forwards, but it is only the horizontal displacement from the edge of the take-off board to the point of landing in the sand pit that counts.

In field events in athletics, the competitors' performances are assessed by the displacements they achieve from a fixed starting position. In the high jump and pole vault the displacements are vertical; but in other events they are horizontal. For this last group of events a trained athlete arrives at the starting position at speed, but must not overstep the position. A long jump competitor (see Fig. 4.1) reaches full speed at the take-off board. Her horizontal speed during the jump is slightly reduced by air resistance or a headwind. However, by jumping high from the board she remains airborne for longer and can thus leap further forward. So a good jump needs both horizontal speed at take-off and height in the air. The flight of a javelin, discus, or shot also requires these two conditions.

Figure 4.1   An international long jump competitor in action
(Mark Shearman)

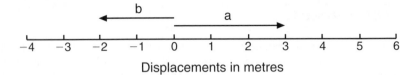

Displacements in metres

Figure 4.2 (a) Positive and (b) negative displacements can be measured from a fixed point

If we extend the scale of a tape measure in the reverse direction from the zero at its end, we can number the scale in the reverse direction with negative numbers as in Fig. 4.2. This means that we can have negative displacements as well as positive ones. We can also change a displacement by adding another one to it. So, for instance, a displacement of three metres added to one of two metres in the same direction makes a final displacement of five metres; and a displacement of minus three metres added to one of two metres makes a final displacement of minus one metre. We can represent displacements in length and direction by lines along the scale as in Fig. 4.3. **A displacement along a line is a vector**, having both a size and a direction.

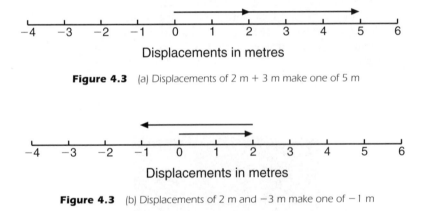

Displacements in metres

Figure 4.3 (a) Displacements of 2 m + 3 m make one of 5 m

Displacements in metres

Figure 4.3 (b) Displacements of 2 m and −3 m make one of −1 m

4.3 When are distances different from displacements?

When competitors in a long distance race complete one lap of the track, they have run 400 m. This is the distance gone. Their displacement from their starting position however is zero, as they are just back where they started. To find a displacement you need to know the initial and final positions. But to find a distance gone you need to know what happened in between. This difference comes about when the athletics track is not a straight line, or when a swimmer swims more than one length of the pool. In the 100 m sprint an athlete's displacement along the track is the same size as the distance gone, because the athlete runs in a straight line. In general

distance gone (m) = average speed (m/s) × time taken (s)   (4.2)

Give an example when a swimmer's displacement from the end of the pool is not the same as the distance travelled.

4.4 Using speed and time to measure distance

A buildings' surveyor can measure the width of a room using an electronic timer and a pulsed high frequency sound emitter. The timer measures how long it takes each sound pulse to cross the room and return as an echo. Inside the device a calculator works out how far sound can travel in this interval, applying Equation 4.2. The distance travelled must be divided by two as the sound crossed the room twice during the timing. So the surveyor's meter shows the width of the room which is:

width (m) = speed of sound in air (m/s) × time taken (s)/2

> **Guinness book of records?**
> Brad Faxon, the American golfer, flew on a Concorde airliner from New York to Spain with team mates for the 1997 Ryder Cup competition. Brad became the first man to sink a putt the full 120 foot length of the passenger cabin.
>
> The aircraft's Captain Jock Lowe, a member of the Concorde Golfing Society, who had set the challenge, also timed the putt. The ball rolled for 23 s from being struck by the putter until it reached the target, a plastic cup lying on its side.
>
> At this time the aircraft was flying at 1330 miles per hour. Captain Lowe calculated that the ball had travelled eight and a half miles from the putter to the cup. Was this the longest golf putt in history? Check it out. (Note: 23 s = 23/3600 hrs).

4.5 Graphs of displacement against time

Fig. 4.4 shows a typical travel graph which has three stages. In the first stage (AB), the traveller takes 30 minutes to travel 2 km. At this rate, the displacement would be 4 km in an hour. So we can see that the velocity was 4 km/h, which is the gradient of the line AB.

In the second part of the graph (BC), the traveller remained at rest at the 2 km position for an hour. The gradient of the line BC is zero, showing that the velocity was also zero.

In the last part of the graph, the traveller returned to the starting position in 20 minutes. The displacement is reduced by 2 km, and the gradient of the graph is

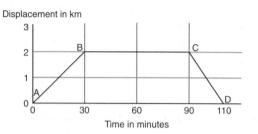

Figure 4.4 A travel graph in three stages

negative, showing that the velocity was also in the negative direction. The gradient of the line CD is: change in displacement/time taken = −2 km/20 min = −6 km/h.

Again the **gradient of the displacement against time graph gives the velocity of the traveller.** This is a general result.

Self test

Fig. 4.5 shows that Anne and Bob set out together to travel 5 km.

(a) Which one arrived first?
(b) What was Anne's velocity?
(c) What was Bob's velocity?
(d) Check that your answers to (b) and (c) above are the gradients of the graphs.
(e) Both Anne and Bob passed Chris. In which direction was Chris travelling?
(f) What was Chris's velocity?
(g) Where was Bob when he passed Chris?

(Answers: (a) Anne, (b) 5 km/h, (c) 2 km/h, (e) Chris and Bob travelled in the opposite directions, (f) −2 km/h, (g) Bob had travelled half way, i.e 2.5 km when he passed Chris).

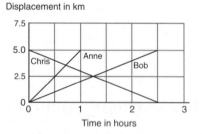

Figure 4.5 Three travel graphs

4.6 Measuring velocity

A trolley speeds up as it runs down an incline. We may want to know its velocity at a certain place, where a narrow beam of light across the path of the trolley falls onto a light sensor. The sensor is an electronic switch which turns on only when bright light

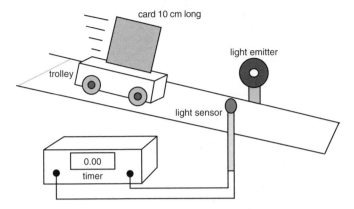

Figure 4.6 As the trolley passes the light emitter, its beam towards the sensor is blocked. The timer records the interval when the beam is interrupted

falls on it. Otherwise it switches off. The sensor is connected to an electronic timer which will only work when the sensor is switched off. A card perhaps 10 cms long is attached to the side of the trolley so that it will interrupt the light beam as the trolley passes. Fig. 4.6 shows this arrangement.

When the trolley is released the timer reading is zero. The card interrupts the light beam and the timer records this interval. It tells how long the trolley took to run forward 10 cms. If this was 0.20 s, we can use Equation 4.1 to find the velocity as follows:

$$\text{Velocity of trolley} = \frac{\text{distance gone}}{\text{time taken}} = 10 \text{ cm/}0.20 \text{ s} = 50 \text{ cm/s} = 0.5 \text{ m/s}.$$

4.7 Graphs of velocity against time

Graphs of displacement against time are not always straight lines. Fig. 4.7a shows the graph for a tennis ball hit vertically upwards. It is tempting to think that the curve shows that the ball travels in a curve, but the ball travels straight up and straight down again. The gradient of the displacement against time graph gives the velocity. Here we can see that the upward velocity reduces to zero in one second, before the velocity becomes negative as the ball falls. After two seconds it is back where it started, having travelled straight up and straight down. What was the highest point it reached?

The velocities of the ball in Fig. 4.7a are found from the gradients of tangents to the curve at different points. They are plotted against time in Fig. 4.7b. The line shows that the upward velocity of 10 m/s reduced to zero in one second and became a downward (negative) velocity as the ball fell. The graph is a continuous straight line.

Fig. 4.8 shows a constant velocity of 3 km/h for a walker who walked for two hours. The shaded area represents the product of the sides of the rectangle, i.e 3 km/h × 2 h. This equals 6 km. The product velocity × time gives the displacement, and it is a general result that **the area between the line and the time axis in a velocity against time graph gives the displacement**.

In Fig. 4.7b we can check this result. The area of the triangle above the time axis from $t = 0$ to $t = 1$ s, is:

$$(1/2)(10 \text{ m/s}) \times (1 \text{ s}) = 5 \text{ m}.$$

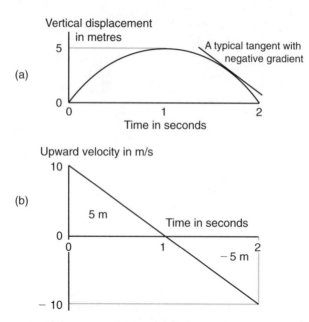

(a)

Vertical displacement in metres

A typical tangent with negative gradient

Time in seconds

(b)

Upward velocity in m/s

5 m

Time in seconds

−5 m

Figure 4.7   Both graphs refer to the tennis ball. The gradients of tangents to the displacement against time graph in Fig. 4.7a are plotted in Fig. 4.7b

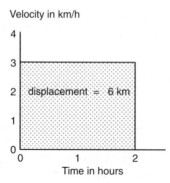

Velocity in km/h

displacement = 6 km

Time in hours

Figure 4.8   The area under a velocity against time graph is the displacement 3 km/h × 2h = 6 km

What did we find was the highest point reached by the ball? Look again at Fig. 4.7a.

In Fig. 4.7b the velocity becomes negative after $t = 1$ s. The triangle below the time axis shows a displacement in the negative direction. The area of this triangle represents −5 m. So first the tennis ball rose 5 m in 1 s, and then it fell the same distance in the next second.

Self test

Fig. 4.9 shows the velocity against time graphs for two cyclists until they came to a halt. Show that Dave travelled twice as far as Kate.

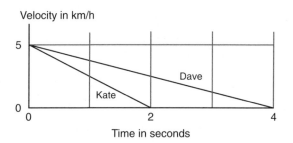

Figure 4.9 *Two cyclists apply the brakes*

4.8 What is acceleration?

The gradients of the graphs for Kate and Dave in Fig. 4.9 show how quickly their velocities were changing. **Acceleration is the rate of increase of velocity**, or:

$$\text{Acceleration} = \frac{\text{increase in velocity}}{\text{time taken}} \tag{4.3}$$

This is the **gradient of a velocity against time graph**. Acceleration is another vector quantity. In Fig. 4.9 Kate's acceleration was: $-(5 \text{ m/s})/2 \text{ s} = -2.5 \text{ m/s/s}$, and Dave's acceleration was: $-(5 \text{ m/s})/4 \text{ s} = -1.25 \text{ m/s/s}$. Note the **units of acceleration are m/s/s, or m/s²** when a **velocity in m/s** is divided by a **time in s**.

A negative acceleration means a deceleration, or an acceleration in the negative direction. Fig. 4.7b has a constant negative gradient for the tennis ball showing a downward acceleration of 10 m/s² for both upward and downward movement of the ball. It may at first seem unlikely, but a stone, a pencil and a tennis ball all fall to the ground with a downward acceleration of about 10 m/s². We shall return to this in chapter 7. For now we should learn that a **constant acceleration** means that the **velocity against time graph will be a straight line**. Appendix A, page 280, has more information about constant acceleration equations.

4.9 Measuring acceleration

Fig. 4.6 shows an experiment in which velocity was measured. An additional stopwatch is needed if we are to measure the trolley's acceleration. The stopwatch is used to time the trolley from rest until its 10 cm card is midway through the light beam. It tells us how long the trolley takes to gain a known velocity. We calculate the acceleration as:

$$\text{acceleration} = \frac{\text{gain in velocity}}{\text{time taken}}.$$

> At the end of the UK Highway Code is some information about the shortest possible stopping distance for a car in an emergency stop on a dry road surface. First, data is given for the 'thinking distance' for each speed, i.e. the distance the car travels before the brakes are applied, once the danger ahead is seen.

Then the 'braking distance' is shown, i.e. the distance the car travels while the brakes are applied as hard as possible. Table 4.2 shows some of the data.

1 Show that the thinking time is (2/3) s, whatever the initial speed.

2 Find the thinking distance for an initial speed of 80 mph.

3 Show that doubling the initial speed results in four times the braking distance.

4 Find the braking distance for an initial speed of 80 mph, and show that the total stopping distance is 120 m.

5 Show that the deceleration during braking is the same for each initial speed.

Table 4.2

Initial speed		Thinking distance	Braking distance	Total stopping distance	Braking time	Deceleration
mph	m/s	m	m	m	s	m/s/s
20	9.0	6	6	12	1.33	9.0/1.33
30	13.5	9	14	23	2.0	
40	18.0	12	24	36	2.67	
60	27.0	18	55	73	4.0	
70	31.5	21	75	96	4.67	

4.10 Circular motion

Motor car engines are sometimes tested by mounting the wheels of the car on rollers, so that the wheels can rotate without the car moving forward. The same thing would happen if the driving wheels of the car were raised clear of the road before the driver tried to drive off. On the rim of the wheel each point moves in a circle. Motion in a circle is a common feature of machine parts.

Fig. 4.10 shows points on the rim of a rotating disc. Each point has the same speed, but as the direction of motion is changing continuously, the velocity is not constant. The speed of the rim is the distance it moves per second, and this is the length of

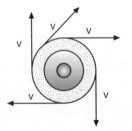

Figure 4.10 Each point on the rim of a rotating wheel moves at speed v, but the points all move with different velocities as shown by the arrows

circumference for one rotation multiplied by the number of rotations in one second, or frequency, f. For a disc of radius r metres:

$$\text{speed at the rim} = \text{circumference} \times \text{frequency} = 2\pi r \times f \, \text{m/s} \qquad (4.4)$$

Each point in a rotating disc has the same frequency which is measured in hertz. One hertz is the same as one rotation per second.

An object moving at uniform rate in circular motion is being pulled towards the centre of the circle. Think about a conker on a string. Without the tension in the string, the conker would fly off along a tangent to the circle. And the faster its speed around the circle, the greater is the force pulling it towards the centre of the circle.

Questions

1 A cycle ride of 20 miles took 2 hours. What was the average speed in miles per hour?
2 Make velocity against time graphs for the journeys of Anne, Bob and Chris in Fig. 4.5.
3 A radar signal takes 2.5 seconds to travel from the Earth's surface to the Moon and back, moving at 300 000 000 m/s. How far is it to the Moon?
4 A swimmer completes (a) 1.5 lengths, (b) 2 lengths and (c) 2.5 lengths of a four length race. Write down her displacements in lengths from the start of the race.
5 In question 4, suppose the pool is 50 m long. How far has she has travelled in each instance?
6 A runner runs 400 m in 60 s. What was his average speed?
7 Show that if you walk at 1 m/s constantly for 1 hour, you will travel 3.6 km.
8 On the same axes make a distance against time graph for an object moving at (a) 1 m/s, and (b) 3 m/s.
9 In question 8, if the objects begin to move at the same time, from the same place, and in the same direction, how far apart will they be after (a) 5 s, and (b) 10 s?
10 An object speeded up from 10 m/s to 25 m/s in 3 s. What was its average acceleration?

What have you learnt?

1 Do you know what a vector quantity is?
2 Do you know that displacements, velocities and accelerations are all vector quantities?
3 Do you know that athletics field events involve distance measurements?
4 Can you describe a journey from a graph of displacement against time?
5 Can you find a velocity from a displacement against time graph, and an acceleration from a velocity against time graph?
6 Do you know how to operate a light beam timer to find a velocity?
7 Do you know what is represented by the area between the line on a graph of velocity against time and the time axis?
8 Do you know how you can measure an acceleration?
9 Do you know what a negative acceleration means?
10 Do you know the difference between a distance travelled and a displacement from a starting position?
11 Can you describe uniform circular motion?

5 | Force and motion

Objectives

After reading this chapter you should
- know that an applied force can accelerate a mass
- know that with no applied force a mass does not accelerate
- be able to find the resultant of several applied forces
- know what forces result from static and dynamic friction
- know what a friction compensated slope is
- understand Newton's second law of motion
- understand that forces are measured in newtons
- be able to calculate the moment of a force about a point
- understand the principle of moments and equilibrium.

5.1 Causing motion

In order to move things around, we must exert forces on them. A force is a vector quantity having a direction as well as a size. And in order to move a stationary object in a given direction, we must apply a force to it **in that direction**. For example you would not push a bobsleigh downhill if you wanted to get it up to the hilltop.

5.2 Motion when there is no applied force

If we exert no force on an object, it displays a property called **inertia**. That means it goes on doing what it has been doing. **If it was already at rest, a mass will remain at rest. But if it was moving at a certain speed in a given direction, it will go on moving with the same velocity until some external source exerts a force on it.** This is Newton's first law of motion.

Sometimes an object slows down when an invisible force of friction acts on it; and sometimes forces act on an object, but it does not change its velocity. This is because the combined effect, or **resultant**, of all the forces is zero. We need to know how to find the resultant of all the forces, taking into account their directions as well as their sizes. The process is called vector addition.

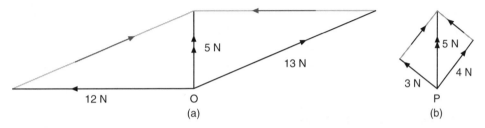

(a)   (b)

Figure 5.1   (a) Forces of 12 N and 13 N acting at O combine to make a resultant of 5 N upwards. In Fig. 5.1b, forces of 3 N and 4 N acting at P also combine to make a resultant of 5 N upwards. The resultant is shown in size and direction by the diagonal of the parallelogram through the point P

5.3 Vector addition

In chapter 4, vectors acting along a line were combined by using an arrow line for each vector. The length of each arrow represented the size of a vector quantity. In this section, the method is extended to combine vectors acting in a plane. The direction of each arrow shows the direction of a force acting at a point.

A force of 5 N could be the resultant of many pairs of forces. Suppose we represent this force by a line 5 units long in a direction shown by the double arrows in Fig. 5.1. The line is one side of a triangle that can be drawn in an infinite number of ways by adding two more sides. Fig. 5.1 shows some examples. The added lines represent by their lengths and directions two forces which combine to produce the resultant of 5 N. Diagrams like these are helpful in adding vectors like forces, velocities or displacements.

In producing a scale diagram to add up two vectors, it does not matter which of the components you draw first. The resultant force of 5 N is the diagonal of a parallelogram passing through the point where the two component forces cross. The resultant has exactly the same effect as the two component forces acting together.

Having added two forces by scale drawing and measurement, we may add a third force and any number of forces by repeating the process of vector addition (as in Fig. 5.2).

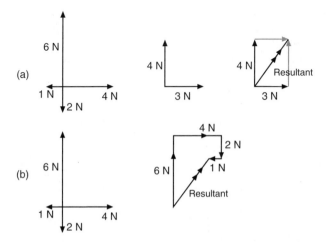

Figure 5.2   (a) Two opposing pairs of forces are added as two forces. In (b) the forces are added singly to give the same resultant

Show that the resultant of two forces of 6 N acting at 120° to one another, is also a force of 6 N, and find its direction.

5.4 Resolving forces

This is the reverse of adding forces by vector addition. A force can be considered the sum of **two component forces which act at right angles to each other**. Fig. 5.3

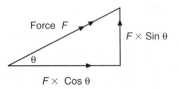

Figure 5.3 F is resolved into components in two perpendicular directions. $F \times \text{Cos } \theta$ is the component of F in a direction making angle θ with F

shows a force F, resolved in the two perpendicular directions represented by lines in the figure. The length of each line shows the size of the component. In general the component along a direction making angle θ with the force F, is $F \times \text{Cos } \theta$

Self test

Show that a force of 40 N resolved in a direction at 60° to the force is a component of 20 N (Use a calculator or scale drawing).

Worked example

Fig. 5.4 is a diagram of a cable-stayed bridge, in which the weight of the roadway is supported by tensions in the attached cables. The weight of the roadway is transferred through the cables to the towers.

Suppose the section of roadway XY is supported by twelve cables joined to the

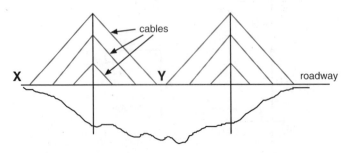
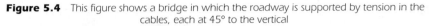

Figure 5.4 This figure shows a bridge in which the roadway is supported by tension in the cables, each at 45° to the vertical

left hand tower (6 cables on each side of the road). Suppose also that each cable makes an angle of 45° with the vertical, and each has the same tension T newtons. The total upward force produced by the 12 cable tensions on the roadway between X and Y is:

$$12 \times T \times \cos 45° = 8.5 \times T \text{ (N)}$$

This upward force must balance the downward force of the weight of the bridge between X and Y. When the bridge becomes loaded with traffic, the roadway sinks slightly and the cables are slightly stretched, so that their tensions increase until the new upward force supports the new weight of the roadway plus its traffic.

Fig. 5.5 shows the Second Severn Crossing bridge under construction. The cable-stayed bridge carries road traffic across the Severn estuary between England and South Wales. Each tower supports 120 cables, (sixty on each side of the road); but the cables make different angles with the vertical, and they have different tensions. This makes it more difficult than in the previous example to calculate the total upward vertical force supporting the road. However, the general principle is the same: **the total upward force produced by the cable tensions equals the weight of the roadway and its traffic**.

The unloaded roadway between the towers, having a mass of 16 000 tonnes, is supported on 120 cables (four groups of 30 cables). Traffic on the bridge does, of course, increase the tensions in the cables, but they must not be stretched beyond their elastic limits (see chapter 2).

Figure 5.5 The Second Severn Crossing Bridge under construction (Haydn Jones)

5.5 Friction

Friction is a force tending to oppose the relative motion of two surfaces in contact as they slide over one another. Often friction is a force just big enough to cancel out a force applied to an object, e.g. when you push a sledge. This has the effect of preventing any change in the object's motion. It may be at rest or it may be moving at a constant velocity. However, if the applied force is greater than the opposing force of friction, the object will accelerate in the direction of their resultant.

The maximum force of friction between two surfaces, called **limiting friction**, depends on the roughness of the surfaces in contact, and the perpendicular force pressing the surfaces together. An empty box may be pushed across the floor with a force which just overcomes the limiting friction as in Fig. 5.6a.

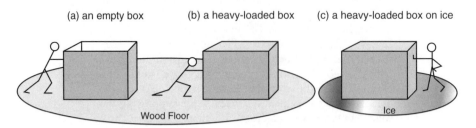

Figure 5.6 This figure shows a box being moved horizontally by a force just big enough to overcome the limiting friction. In (a) the limiting friction is small as the empty box is not heavy. In (b) the box is heavy, and the limiting friction is proportionally greater. In (c) the box is heavy, but the nature of the surfaces in contact makes the limiting friction very small

If, however, the box is filled with books or bricks, a greater force is needed to push it across the floor, because the limiting friction has increased. **When an object is stationary, the limiting friction increases with the component of force pressing the surfaces together**.

The heavy box full of bricks can be slid quite easily across ice, which shows that the limiting friction depends also on the surfaces in contact (see Fig. 5.6c). The runners of a bobsleigh may be waxed in order to reduce its limiting friction on ice. Friction in machines is often reduced by means of roller bearings or ball bearings. A thin oil film between the moving parts of a machine may hold sliding surfaces apart to reduce friction, increase efficiency and prevent rapid wear.

5.6 Friction compensated slopes

A trolley will run down a steep incline accelerating as it goes. On a very gentle gradient it may run down at constant velocity if given an initial push to get started. This gentle incline is called a **friction compensated slope**, because the trolley in motion is subject to no resultant force. However, it is only friction compensated for the trolley moving down the slope. If it were moving up the slope the trolley would slow down because there would be a resultant force on it. Friction opposing the motion would act down the slope.

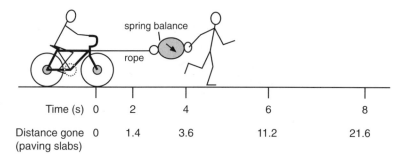

Time (s)	0	2	4	6	8
Distance gone (paving slabs)	0	1.4	3.6	11.2	21.6

Figure 5.7 *A constant forward force accelerates the bicycle. We can estimate the speed after 7 s by finding the average speed from 6 s to 8 s after the start. i.e. estimated speed at t = 7 s is*
$$v = (21.6 - 11.2) \text{ slabs}/2 \text{ s} = 5.2 \text{ paving slabs/s}$$

5.7 Accelerated motion

Here is an outdoor experiment needing 5 people, a bicycle, a stopwatch, about 3 m of rope, and a spring balance able to weigh masses of up to about 5 kg. The units of force do not matter for this experiment, which is conducted on a level pavement having paving slabs of equal width so that the distances gone can be measured by counting the paving slabs.

Fig. 5.7 illustrates a rider free-wheeling along the pavement. The bicycle is pulled by a rope kept at constant tension by a runner holding the spring balance. The stopwatch is started as the force is applied and the bicycle starts to move forward. Every two seconds the time is called out, and a stone is placed on the pavement opposite the front wheel of the bicycle. From the positions of the stones it is possible to see that the cycle travelled further in each interval of 2 s than in the one before. It was speeding up, or accelerating, at a fairly steady rate. A series of tests reveals that the acceleration is bigger when more forward force is applied; but the data is not easily repeatable in this experiment, as it is difficult for the runner to keep the spring balance reading constant while on the move.

A more accurate experiment can be carried out in the laboratory. If a trolley on a friction compensated slope experiences no resultant force, we may cause its acceleration by **applying an extra force** down the slope. Immediately we can see that the trolley picks up speed down the slope. **The acceleration is in the same direction as the resultant applied force**.

Fig. 5.8 shows an experiment using the slope to see how resultant force and mass

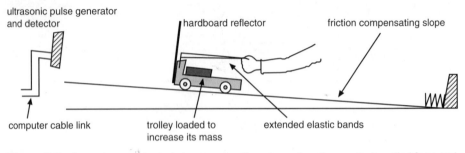

ultrasonic pulse generator
and detector

hardboard reflector

friction compensating slope

computer cable link

trolley loaded to
increase its mass

extended elastic bands

Figure 5.8 Apparatus used to study how accelerations depend on the resultant applied force and mass of a trolley

affect the trolley's acceleration. Forces are applied by means of one or more similar elastic bands attached to the trolley and extended a fixed distance. Each elastic band exerts the same force on the trolley, so the force can be written down as one, two, or three units, etc. according to the number of elastic bands we are using.

A hardboard reflector is attached to the back of the trolley. This reflects ultrasonic waves emitted by a pulse generator at the top of the slope. The distance between the generator and the reflector can be found as we described in Section 4.5. In this experiment the measurements are fed to a computer every 0.05 s, and a graph of distance against time, like Fig. 5.9a, is shown on the computer screen as the trolley accelerates down the slope.

Notice that in Fig. 5.9a the distance against time curve gets steeper as the trolley speeds up. You may also see the effect of the trolley rebounding up the slope from a stiff spring at the bottom. The trolley's velocity at each moment is measured by the gradient of the curve, and the computer programme enables the velocities to be displayed as in Fig. 5.9(b).

The first part of the velocity against time graph is a straight line. This shows that

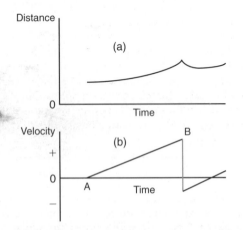

Figure 5.9 (a) The computer screen display shows the distance against time graph for the trolley pulled down the slope by a constant force. (b) displays the velocities from the gradients of the first graph. The straight sections show that the velocity increased uniformly, and the gradient of the section AB gives a measurement of the acceleration in arbitrary units

Table 5.1 The accelerations of a trolley on a friction compensated slope (the units are arbitrary)

Number of elastic bands ∝ Applied force	Trolley mass	Acceleration	Force/mass
1	1	0.5	1
2	1	1.1	2
3	1	1.5	3
4	1	1.9	4
4	2	1.0	2
4	3	0.7	1.3
4	4	0.5	1

the acceleration down the slope is constant. Its value can be found from the gradient of this part of the line.

The experiment is repeated and the trolley's acceleration is recorded for forces of one, two, and three units, etc. Later the same force is applied to trolleys whose mass has been doubled, then trebled, etc. by the addition of equal trolley masses. The accelerations are recorded in Table 5.1.

5.8 Analysing the data in Table 5.1

1   The effect of increasing the force has been to increase the acceleration, as shown in the first four tests. More accurate experiments have established that **the acceleration is directly proportional to the resultant applied force.** In symbols:

$$a = \text{constant} \times F \tag{5.1}$$

2   The effect of increasing the mass for the same resultant force was to decrease the acceleration as shown in the last four tests. Further experiments have established that the acceleration is proportional to 1/mass. In other words, **the acceleration is inversely proportional to the mass**. In symbols:

$$a = \text{constant}/m \tag{5.2}$$

Combining all our results, **the acceleration of the mass is in the direction of the resultant applied force, and it is proportional to the resultant applied force/mass**. In symbols:

$$a = \text{constant} \times F/m \tag{5.3}$$

This is Newton's second law of motion.

5.9 Defining the force of one newton

We define **one newton as the resultant force which causes a mass of one kilogram to accelerate at one metre per second per second**. The newton is the unit of force, in a system based on the metre, kilogram and second. Having defined its size, we can write Newton's second law of motion in the form:

$$a = F/m, \text{ or } F = m \times a \tag{5.4}$$

where F is the resultant applied force in newtons, m the mass in kilograms, and a the acceleration in metres per second per second.

Example
Find the acceleration of a trolley of mass 1.5 kg pulled across a horizontal table top with a force of 4 N against a friction force of 1 N.

Answer
In symbols $a = F/m$. The resultant force $F = 4\,\text{N} - 1\text{N} = 3\,\text{N}$, so a $= 3\,\text{N}/1.5\,\text{kg} = 2\,\text{m/s}^2$ in the direction of the 4 N force.

Example
What uniform braking force will stop a car of mass one tonne, i.e. 1000 kg, travelling at 20 m/s in 8 s?

Answer
In symbols $F = m \times a$, where $m = 1000$ kg and $a = -20/8\,\text{m/s}^2$. So $a = -2.5\,\text{m/s}^2$, and $F = 1000\,\text{kg}(-2.5\,\text{m/s}^2) = -2500\,\text{N}$. The braking force is 2500 N. (Slowing down means a is negative).

Example
(a) Draw a velocity against time graph for the car in the last example.
(b) Estimate the area between the graph and the time axis.
(c) How far did the car travel while the brakes were applied?
(d) How much work in joules was done against the braking force?
(e) Where did the energy come from to do this work?

Answers
(a) See Fig. 5.10.
(b) The area = half base × height = $(\frac{1}{2})8\text{s} \times 20\,\text{m/s}$ = 80 m.
(c) The area = distance gone = 80 m.
(d) Work done = force × distance = 2500 N × 80 m = 200 000 J
(e) The car's loss of KE equals the work done.

Self test

Show that a mass of 10 kg experiencing a resultant force of 40 N will gain a velocity of 8 m/s from rest in 2 s.

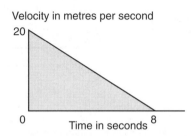

Velocity in metres per second

Figure 5.10 *The shaded area beneath the line shows the distance the car travelled*

Example of an airliner taking off
A typical take-off mass for a 747 airliner is 340 000 kg, depending on the total load of passengers, freight, and fuel on board. The take-off speed depends on the air pressure and temperature, but a typical speed at rotation (at which the pilot rotates the plane for take-off) is 75 m/s. The four Rolls-Royce RB-211 engines produce a total thrust of 952 000 N on average at the low speeds up to take-off.

Estimate the length of the take-off run in still-air before rotation begins. The aircraft's acceleration is F/m, the ratio of the thrust to the mass.
This is: $a = 952\ 000$ N$/340\ 000$ kg $= 2.8$ m/s^2.
The time to reach 75 m/s from rest is $t = (75/2.8)$ s $= 27$ s. During the take-off run the average speed is 0.5×75 m/s $= 37.5$ m/s.
The distance gone equals
average speed $\times$ time $= 37.5 \times 27 = 1010$ m.

Further food for thought
How will a headwind affect the length of the take-off run?
How will conditions differ when the aircraft is landing?

5.10 Rotation

A spanner causes rotation when doing up a nut on a bolt as in Fig. 5.11. A longer spanner gives more leverage, and a bigger turning effect is achieved with the same effort. The turning effect also depends on the component of force perpendicular to the length of the spanner. If the force is reversed, rotation in the opposite sense is achieved, and the nut is undone. The two rotations are distinguished by calling them *clockwise* and *anticlockwise*.

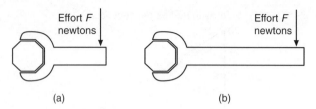

Figure 5.11   An effort applied further from the axis of rotation has a greater turning effect. So the longer spanner is more effective

5.11 Moment of a force about a point

The turning effect of a force F about a pivot P, (see Fig. 5.12) is the force multiplied by the perpendicular distance, d, between the force and the pivot. **$F \times d$ is called the clockwise moment of the force about P**, because it tends to rotate the bar in a clockwise sense.

In Fig. 5.13(a) another force is applied, giving a clockwise moment of $F' \times d'$.

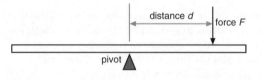

Figure 5.12   The moment of the force F about the pivot is $F \times d$ clockwise

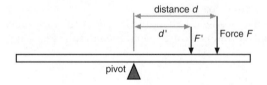
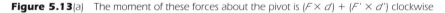

Figure 5.13(a)   The moment of these forces about the pivot is $(F \times d) + (F' \times d')$ clockwise

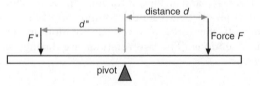

Figure 5.13(b)   The moment of these forces about the pivot is $(F \times d) - (F'' \times d'')$ clockwise, as the force F'' has an anticlockwise moment

The total clockwise moment becomes the sum of the individual ones, i.e. $(F \times d) + (F' \times d')$.

An anticlockwise moment is produced by exerting a force on a bar in the opposite sense. In Fig. 5.13(b) the total clockwise moment about the pivot is $(F \times d) - (F'' \times d'')$, because F'' has an anticlockwise moment. The turning effect or moment of a force about a point is also called its **torque** (pronounced as 'talk').

5.12 Equilibrium

If an object remains at rest while acted upon by a number of forces in one plane, the resultant force on it must be zero. But also the turning effects must balance. So for an object at rest, **the sum of the clockwise moments of all the forces about any point, must equal the sum of the anticlockwise moments about this point**. This is the **principle of moments** which applies whenever there is no tendency for an object to rotate.

— **Self test** —

A see-saw is balanced on a pivot at its mid point. Show that a child weighing 400 N sitting 2.5 m from the mid point could be balanced by another weighing 500 N sitting 2 m from the mid point on the other side.

— **Questions** —

1 Forces of 12 N and 16 N at right angles act on an object. Find their resultant by vector addition.

2 A car with its brake off runs down a slope at a constant 0.5 m/s. Why does it not speed up?

3 (a) From Table 5.1 make a graph of the acceleration plotted against the force/mass.
   (b) What tells you that the trolley's acceleration was proportional to the force/mass?
   (c) If you applied twice as much force and doubled the mass, what would happen to the acceleration of the trolley?

4 Two cars are of identical design. Their engines are identical, but one has steel bodywork and the other has aluminium. Explain which car can accelerate faster.

5 Cars with ABS stop without skidding. This can mean stopping in a shorter distance. What does this tell you about limiting static friction compared with sliding or dynamic friction?

6 (a) What resultant force accelerates a car of mass 800 kg uniformly from 0 to 60 mph in 9 s? (Take 60 mph as 27 m/s).
   (b) What force will slow the car from 60 mph to 30 mph in 4.5 s?

7 Table 5.2 shows data for the take-off run of a Boeing 777 airliner of total mass 262 tonnes (i.e. 2.62×10^5 kg). The aircraft left the ground at an airspeed of about 78 m/s. A stopwatch was used, and the airspeed was recorded at different times during the aircraft's take-off run. Airspeed is the speed of the aircraft through the air. This may differ from the aircraft's ground speed if a wind is blowing. (We did not borrow an airliner to obtain

Table 5.2

Aircraft's airspeed (m/s)	24	36	41	49	60	71	80	90
Stopwatch time (s)	5	10	12	15	20	25	30	35

Table 5.2. A flight simulator on the ground was used instead!).
(a) Plot the data as a graph of airspeed against time.
(b) Use the graph to find the aircraft's acceleration from $t = 5$ s to $t = 20$ s.
(c) Calculate the net force accelerating the aircraft along the runway.
(d) Give reasons why the acceleration of an aircraft may change after take-off.
(e) Estimate the distance to take-off for this aircraft in still-air conditions.
(f) What advantages are there in taking off into a headwind?

8 A given engine thrust causes a car to accelerate, but the acceleration is smaller when the car is travelling at speed than when it is slow moving. Explain this.

9 In icy conditions a farmer puts a heavy sack in the boot (or trunk) of his car over the rear axle. Explain why this can improve the car's handling.

10 Commercial scales support a load on a horizontal bar 20 cm from the pivot. A counter-weight of 25 N on the opposite arm of the bar can be moved up to 80 cm from the pivot (see Fig. 5.14). What is the weight of the maximum load that can be weighed by balancing the bar?

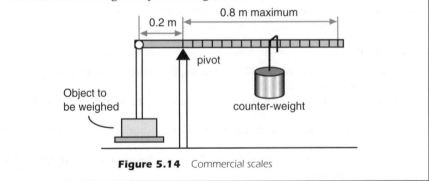

Figure 5.14 Commercial scales

What have you learnt?

1 Do you know what it means when we say a mass has inertia?
2 Do you know how to find the resultant of two forces by vector addition?
3 Can you find the component of a force in a given direction?
4 Do you know what effects the limiting static friction between two surfaces?
5 Do you know what a friction compensated slope is?
6 Can you explain how to find the acceleration of a mass?
7 Do you understand Newton's first and second laws of motion?
8 Do you know how one newton of force is defined?
9 Do you understand the moment of a force about a point?
10 Can you apply the principle of moments?

6 Work, energy, power and momentum

6.1 Work and energy

When an object is accelerated by a force, work is transformed into kinetic energy. Question 6 on page 47 illustrated this process in reverse, when the kinetic energy of a car was transformed into work done against the force of the brakes. Energy is measured by the amount of work it can do, measured by a force × distance it moves something along its line of action.

On this basis we can identify other forms of energy. Heat, for example, applied to gas in a cylinder, causes its pressure to rise, so that it can do work pushing a piston along the cylinder. Heat can be derived from many sources including light, chemical energy, sound, rotational, electrical and nuclear energies. Energy can take many other forms, including wave energy, elastic spring energy and the potential energies of certain things in gravitational, electrical or magnetic fields. All of these can give rise to forces which may do some work.

6.2 Law of conservation of energy

This universal law applies through all science and technology. It has been described as the most fundamental law of nature, and it has no special conditions to limit its application. The law states that **energy cannot be created or destroyed, but it can be changed from one form into another**.

6.3 Transforming energy

Inventors and designers have devised many ways of transforming energy. A loudspeaker, for example, converts electrical energy into sound waves, and a microphone is used for the reverse change. An electric motor transforms electrical energy into rotational energy, and an a.c. generator can reverse that change. A car converts chemical energy of a fuel into KE, etc. However, none of these devices or **transducers** which transform energy can output more energy than the amount of energy put into them. They cannot create or destroy energy.

In a power station, either chemical or nuclear energy is transformed into electrical energy. This is carried out in a number of steps, which we represent with boxes.

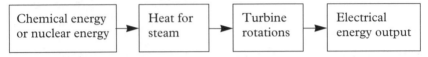

6.4 Efficiency of a transducer

One of the problems of transforming energy is that a transducer seldom converts all the energy into the form you want. Loudspeakers, theatre spotlights and electric motors all generate some heat, which is obviously inefficient. It happens on a larger scale in power stations where only a fraction of the heat, perhaps only 25%, can be transformed into electrical energy. This has given rise to community heating projects in some areas where the heat energy not converted into electrical energy in the power station has been transferred to heat buildings in the locality.

The efficiency of a transducer = useful energy output/total energy input. The fraction is usually multiplied by 100 and given as a percentage.

Hydroelectric power stations use the potential energy of water in reservoirs up in the hills to drive turbines and generate electricity down in the valleys. Steam is not required and so there is negligible heat loss from an HEP station. However not all the potential energy of the water is converted into electrical energy; some remains as kinetic energy of the water leaving the turbines. Efficiencies as high as 80% are an attractive feature of HEP.

6.5 Power

The power of a machine is the rate at which it transforms energy. Power is measured in joules per second or watts (W). One joule per second = one watt. So:

Power = work done/time taken.

Electrical appliances are labelled according to the power input from the mains supply. For example a 15 W filament bulb transforms 15 joules of electrical energy into heat and light energy in one second. A 100 W filament bulb transforms 100 joules per second. Not surprisingly the 100 W bulb is much brighter, and it also gives out more heat. A more efficient 'long life' bulb of power 15 W appears as bright as the 100 W filament bulb and it gives out less heat. The long life bulb costs more than the filament bulb, but it uses less power, lasts longer and, the makers claim, it is more economical.

The energy transformed in t seconds by a device whose power is P watts can be worked out from the equation: **energy** = **power** × **time** = $P \times t$ (J). This is quite an important result because we pay the electricity company for the units of electrical energy we have used. For this purpose a larger unit is chosen called the kilowatt hour (kWh). The same equation applies as above, but the power is in kW, and the time in hours.

Example
If one unit (kWh) costs 8 p, find the cost of using a two kW heater for three hours.

Answer
Energy = power × time = 2 kW × 3 h = 6 kWh.
Total cost = 6 kWh × 8 p/kWh = 48 p.

6.6 Kinetic energy

Fig. 6.1 shows a graph of velocity against time for a vehicle as it gains a velocity v metres per second at a constant rate from a standing start. The distance gone, s, in t seconds is the area between the line and the time axis, which = half base × height or $s = v \times t/2$ metres. The acceleration $a = v/t$, which is the slope of the graph.

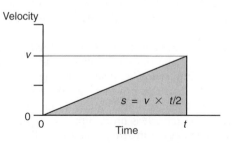

Figure 6.1 A graph of velocity against time showing a velocity, v, gained from rest in time, t. The shaded area shows the distance, s, travelled. The slope of the line, $v/t = a$, the acceleration

The kinetic energy gained by the vehicle is equal to the work done by the resultant force F, where $F = m \times a$. So the KE = $F \times s = m \times a \times s = m(v/t)(v.t/2)$ from above. The t cancels to leave an expression for the KE = $(1/2) m \times v^2$ (6.1)
Energy is a scalar quantity, being defined by its size alone.

Self test

Use Equation 6.1 to check that a car of mass 1000 kg travelling at 20 m/s has KE of 200 000 J = 200 kJ.

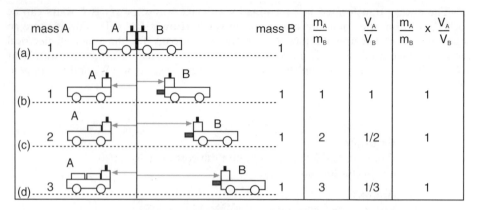

mass A			mass B	$\dfrac{m_A}{m_B}$	$\dfrac{V_A}{V_B}$	$\dfrac{m_A}{m_B} \times \dfrac{V_A}{V_B}$
(a)	1		1			
(b)	1		1	1	1	1
(c)	2		1	2	1/2	1
(d)	3		1	3	1/3	1

Figure 6.2 Trolleys A and B are shown as they recoil from each other after a spring in B is released. The mass of A can be increased by adding lead lumps of equal mass. When the trolleys recoil, they do so with equal momentum, the larger mass moving with a smaller velocity.
Momentum = mass × velocity

6.7 Recoiling trolleys

Two trolleys, A and B, are arranged on a level bench top as in Fig. 6.2a. A compressed spring in trolley B is released so that trolley A is pushed away to the left. The action of pushing on trolley A results in a force on trolley B in the opposite direction. In Fig. 6.2b, the trolleys are shown as they might appear in a photograph taken just after the spring release. The trolleys have moved equal distances in opposite directions, which shows they are travelling at the same speed. The forces on the trolleys must have been equal in size.

In a second test the total mass of trolley A is doubled by adding a lead lump to it. This time trolley B, with half the mass of trolley A, travels twice as far in the same time, so it must be moving twice as fast (see Fig. 6.2c). If we calculate the mass × velocity for each trolley, the products would be the same size, though they would have opposite signs because one velocity is negative. If we add up the two products, the result would be zero. It is more difficult to tell, but the forces on the two trolleys must again have been equal in size, since the trolley with double the mass was accelerated to only half the speed in the same time. Remember $F = m \times a$ for each trolley.

In Fig. 6.2d, the mass of trolley A has been increased to three units by adding another lead lump. The positions of the trolleys shows that trolley B has travelled three times as far as trolley A, so it has gained three times the speed. Again the forces on the trolleys were the same size, and the sum of the products mass × velocity for each trolley comes to zero.

6.8 Law of conservation of momentum

The **momentum of an object** is the product of its **mass × velocity**. It is a vector quantity, having the direction of the object's velocity, and typical units kg × m/s. In these tests the total momentum in the system of two trolleys remained zero throughout all the tests. In a system of freely moving masses, **the total momentum of all the** masses measured in any given direction remains unchanged, provided no external force is applied to the system in that direction. This is called the **law of conservation of momentum**.

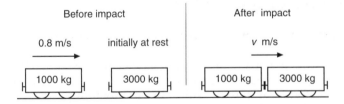

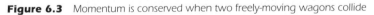

Figure 6.3 *Momentum is conserved when two freely-moving wagons collide*

Conservation of momentum examples occur in a railway marshalling yard and in the explosive separation of a space vehicle and its used-up fuel tanks. Fig. 6.3 shows an impact between two freely-moving goods wagons. Before the collision, the larger mass is at rest and has no momentum. The smaller mass collides with the larger, and the two wagons link together as they move along the track. Momentum is conserved in this collision. Before impact, the wagons have momentum from left to right of:

$$1000 \text{ kg} \times 0.8 \text{ m/s} + 3000 \text{ kg} \times 0 \text{ m/s} = 800 \text{ kg. m/s}$$

The momentum after impact is the same, since momentum is conserved, i.e. $(1000 \text{ kg} + 3000 \text{ kg}) \times v$ m/s = 800 kg. m/s where v m/s is the final speed of the two wagons from left to right. So $4000 \times v = 800$ kg. m/s, and $v = 800/4000 = 1/5$ or 0.2 m/s.

In this example it can be shown that the final kinetic energy of the wagons is less than its initial value. In order to return the system to the initial conditions we need no more momentum, but we should have to increase the energy of the system. This is what happens in space when, by using explosive charges, a separation is made between two parts of a space vehicle. The explosion causes only internal forces which leave the total momentum of the system as a whole is unchanged.

6.9 Internal and external forces

These words distinguish forces which accelerate a mass from those which do not. If I sit in the back seat of my car and push the front seats forwards, the car does not move; but if I get out and push from outside the car, it might move. Forces which might cause a mass to accelerate are applied by an external source, which is separate from the mass. In the recoiling trolley experiments, each trolley experiences an external force. But if we consider the pair of trolleys together, then the forces of the spring are internal to this system.

6.10 Impulse of a force

We know from Newton's second law of motion that an external force = mass × acceleration.

This can be written: Force = mass × gain in velocity/time taken. Getting rid of the fraction, this becomes: Force × time taken = mass × gain in velocity, and this is often written in the form: **impulse = gain in momentum**, where the product **force × time taken is called the impulse of the force**. Impulse, like momentum is a vector quantity.

Car seat belts and air bags
Suppose a car travelling at 30 mph, or nearly 50 km/h, stops suddenly in a collision. Everything in the car will travel straight on, unless a restraining force can be exerted on it.

The purpose of a seat belt is to reduce the maximum force which the driver or any passenger will experience. When the rapid deceleration begins, all the seat belts lock, and anyone wearing a seat belt starts to experience the belt's restraining force. By applying a smaller force for a longer time, the belt can slow down a passenger safely without injury. This is an example where the restraining force F, times the time t for which it acts is equal to the person's loss of momentum. If t is increased, then F can be reduced and the same reduction in momentum will occur.

An air bag has a similar effect to a seat belt, and it is particularly useful for the driver. As the car suddenly stops, the bag rapidly inflates. The driver's head is in danger of hitting the car's steering wheel. However, if it first comes into contact with a gas-filled bag, a smaller force can slow the motion of the head more gradually.

Fig. 6.4 shows a car undergoing a frontal impact test. While the passenger compartment is designed to protect the driver and passengers, the crumple zones at the front and rear of the car are designed to absorb energy as they buckle during impact, leaving the passenger space undistorted. Strengthening bars in the doors of many cars give their occupants added side impact protection.

6.11 Newton's third law of motion

In the trolley experiments above, when trolley A was pushed to the left, trolley B experienced an equal force to the right. Isaac Newton said that **whenever one object**

Figure 6.4 A frontal impact test on a car
(Photo courtesy of the Transport Research Laboratory)

exerts a force on another, the second object exerts an equal force in the opposite direction on the first. This is **Newton's third law of motion**. We rely on this
law for transport. A vehicle is pushed forwards only as it pushes something else backwards by means of it wheels, a propeller, or rocket motors.

6.12 Rocket propulsion

Forward propulsion in a rocket is achieved by thrusting the part of the rocket we call
the spent fuel in the backward direction. This provides an equal forward thrust on the
mass of the rocket. It is like sitting in a dinghy on a lake and throwing out coconuts
over the stern (Fig. 6.5). The forward force produced by this process is equal to the
change of momentum of the 'fuel' per second.

The good news about rocket propulsion is that it works in space where no other
type of propulsion unit is effective, because there is nothing in space to push against.

Figure 6.5 A dinghy loaded with coconuts illustrates how rockets achieve forward propulsion by
ejecting their fuel backwards

The bad news is that once all the fuel is used, no more thrust is available. You don't want to be sitting in the dinghy too far from shore when the coconuts run out, unless you have also taken a pair of oars; sadly, oars are not much use in space.

When a massive craft is sent up into the atmosphere, the rocket motors ignite some time before it leaves the launch pad. For lift-off to occur, the resultant force on the craft must be upwards, so the rocket thrust must build up until it is greater than the total downward weight of the craft.

Self test crossword

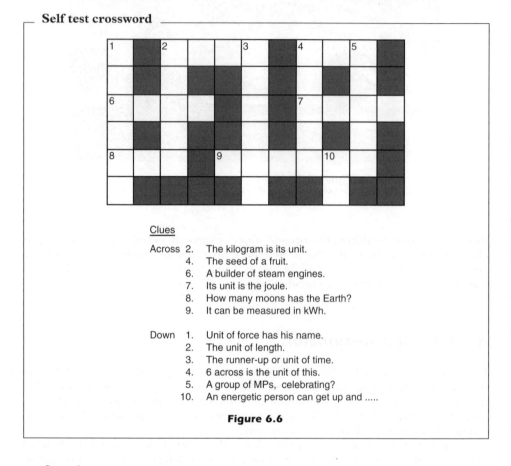

Clues

Across 2. The kilogram is its unit.
4. The seed of a fruit.
6. A builder of steam engines.
7. Its unit is the joule.
8. How many moons has the Earth?
9. It can be measured in kWh.

Down 1. Unit of force has his name.
2. The unit of length.
3. The runner-up or unit of time.
4. 6 across is the unit of this.
5. A group of MPs, celebrating?
10. An energetic person can get up and .....

Figure 6.6

Questions

1 List three forms of energy.
2 A diesel locomotive hauls an inter-city passenger train.
  (a) What form of energy does the fuel possess?
  (b) What useful form of energy is derived from driving the train?
3 In question 2, why is this energy change not 100% efficient?
4 How much KE has a train of mass 200 000 kg travelling at 28 m/s? (Use $KE = m \times v^2/2$).
5 A tennis ball gains 40 J of KE after being in contact with the racket for 0.05 s. What was the power of this energy conversion?
6 One snooker ball collides with another. What two laws of conservation apply to this?

7   In collisions KE is not usually conserved while momentum is. Where does the lost KE go in collisions between motor vehicles?
8   A space-walking astronaut remains attached to his spacecraft by a strong cable. How might the cable be used?
9   How do rockets provide thrust in space?
10 A travelling space capsule is separated from its parent craft by a small explosive charge pushing them apart. This impulse increases the forward momentum of the capsule by 2000 kg.m/s. What is the effect of this separation on the parent craft?

What have you learnt?

1   Do you know that work done = force × distance moved?
2   Do you know how energy and work are related?
3   Can you list several forms of energy?
4   Do you know the meaning of the efficiency of a transducer?
5   Can you apply the law of conservation of energy?
6   Can you find the KE of a mass using the formula $KE = m \times v^2/2$?
7   Do you know that power = work done/time taken, and its units are watts (W)?
8   Do you know how to calculate the momentum of a moving object?
9   Do you know the law of conservation of momentum?
10 Do you know Newton's third law of motion?
11 Can you explain rocket propulsion?

Gravity

After completing this chapter you should
- know the difference between mass and weight
- know how to find the acceleration of a freely falling mass
- understand the force of air resistance
- understand that a falling mass loses potential energy
- understand that a falling mass gains kinetic energy
- understand satellite orbits and planetary orbits.

7.1 A force without contact

An object weighed on a spring balance is in equilibrium under the action of two opposing forces. It experiences an upward force from a stretched spring with which it is in direct contact (see Fig. 7.1). And it experiences the downward force of gravity caused

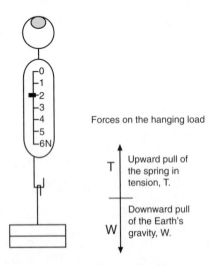

Forces on the hanging load

T — Upward pull of the spring in tension, T.

W — Downward pull of the Earth's gravity, W.

Figure 7.1 A spring balance supports a mass of weight W with an upward force T the same size. The scale reading shows T = 2 N

because the Earth is nearby; yet there is no direct contact between the object and the Earth. Non-contact forces like gravity and the forces between two magnets have been called **action at a distance**.

Any two objects attract each other with a force of gravity, even if they are not in contact. For example, the Earth exerts this force on artificial satellites which move round it in orbits. It exerts this force on the Moon, over a distance of nearly 400 000 km, which results in the Moon staying in orbit round the Earth. And the Sun exerts this force on the planets at even greater distances, so that they remain in orbits round the Sun.

In each of these examples the force of gravity is exerted without any direct contact between the two attracting objects.

7.2 Weight and mass

The force of gravity acting on an object is called its **weight**. This invisible force is registered on a spring balance supporting the object.

In everyday speech 'weight' and 'mass' are often muddled up. The **mass** of an object indicates the amount of material in it. This is measured in kilograms or other units of mass. If an object is moved to another part of the universe, **the mass of the object does not change**, as it still contains the same amount of material.

Weight is measured in newtons. **The weight of an object may change, depending on where it is situated**. If an object has a mass of m kilograms, then its weight is $m \times g$ newtons, where the force of gravity on each kilogram is g newtons. Near the Earth's surface the value of g is about 10 N/kg, so a man of mass 80 kg has a weight on Earth of about $80 \times 10 = 800$ N.

7.3 Falling freely

An object released from rest and allowed to fall freely, has a downward acceleration due to gravity. This is an example of Newton's second law of motion. With no air resistance, the resultant force on the object is $m \times g$ newtons downwards; and as this force is equal to the mass $\times$ acceleration, we may write the equation:

$$F = m \times g = m \times a \tag{7.1}$$

It follows that $g = a$, where a is the free fall acceleration due to gravity. If $g = 10$ N/kg, then $a = 10$ m/s², whatever the mass of the object. This explains why a tennis ball, a pencil and a stone, having different masses, all experience the same initial acceleration in free-fall.

___ Self test ___

Why is the free-fall acceleration of a rock near the Moon's surface much less than it would be near the Earth's surface?

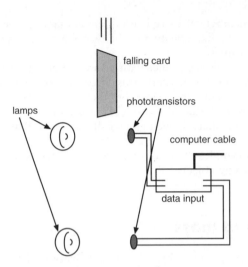

Figure 7.2 The falling card interrupts two light beams. Data fed from phototransistors to a computer enables the card's acceleration to be found

7.4 Measuring free-fall accelerations

Experiments for measuring smaller accelerations were described in earlier chapters. Here are two more methods for finding accelerations in free-fall.

The first method relies on computer software and the fact that computers have very accurate timing circuits. A stiff rectangular card about 15 cm long falling freely, interrupts light beams reaching two phototransistors, arranged one above the other about half a metre apart (see Fig. 7.2). The computer records three time intervals: the time for which each beam was broken, and the interval between the two beams being broken. Once the length of the card is keyed into the computer, the software calculates the average speed of the card at each phototransistor, and finds the acceleration from the **gain in velocity/time taken for the increase to occur.**

The second method uses multiple flash photography. A falling steel ball is photographed repeatedly at regular known intervals. The photograph as in Fig. 7.3 is analysed, and the distance gone between two images is used to find the average velocity in that interval. The acceleration is calculated as in the previous paragraph.

7.5 Air resistance

The force of air resistance depends on the shape and size of an object as well as its speed through the air. Like friction, this force opposes an object's motion. The force of air resistance increases with the speed of the object.

The effect of air resistance is most obvious on a sky diver. After jumping from an aeroplane, a diver accelerates downwards at about 10 m/s^2, because the force of gravity, i.e. weight, is acting downwards. As her downward velocity increases, the upward force of air resistance also builds up, and the resultant downward force, (weight minus air resistance) gets less. So her downward acceleration gets less the faster she falls.

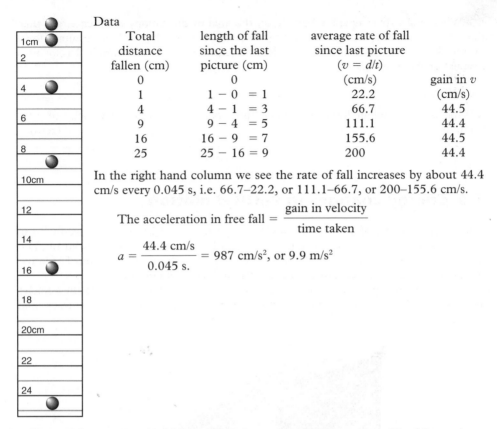

Data

Total distance fallen (cm)	length of fall since the last picture (cm)	average rate of fall since last picture $(v = d/t)$ (cm/s)	gain in v (cm/s)
0	0		
1	1 – 0 = 1	22.2	
4	4 – 1 = 3	66.7	44.5
9	9 – 4 = 5	111.1	44.4
16	16 – 9 = 7	155.6	44.5
25	25 – 16 = 9	200	44.4

In the right hand column we see the rate of fall increases by about 44.4 cm/s every 0.045 s, i.e. 66.7–22.2, or 111.1–66.7, or 200–155.6 cm/s.

$$\text{The acceleration in free fall} = \frac{\text{gain in velocity}}{\text{time taken}}$$

$$a = \frac{44.4 \text{ cm/s}}{0.045 \text{ s.}} = 987 \text{ cm/s}^2, \text{ or } 9.9 \text{ m/s}^2$$

Figure 7.3 A small steel ball falls freely from the rest. The total distance the ball has fallen can be measured on the centimetre scale in the figure. The time between successive positions was 0.045 s. How can you tell the ball was speeding up?

___ **Self test** _____

A good scientist looks for patterns in the data. What numbers would you expect in the next row of the data table above?

Eventually the diver reaches a **terminal velocity** when the air resistance is as large as her weight. The resultant force on her becomes zero and her acceleration becomes zero also. The terminal velocity for a human body is about 200 km/h, but it depends on the shape adopted during the fall (see Fig. 7.4).

Figure 7.4 A falling sky diver

When a sky diver opens a parachute, the area of the canopy is much larger than the area of her body, and the air resistance becomes much larger. The resultant force on the diver is upwards and she decelerates until a new, smaller terminal velocity is reached of about 30 km/h.

Self test

What is the resultant force on a parachutist falling at a terminal velocity of 30 km/h? (zero)

7.6 Energy changes in vertical motion

The **kinetic energy** of a mass, m, moving at speed, v, is $(1/2)\, m \times v^2$ as we found in the last chapter. Clearly free-falling objects gain kinetic energy if they speed up, and this must result from a transformation of energy from some other form. The other form of energy is called **potential energy**, PE, or energy of position.

Consider the fairground ride in Fig. 7.5. At the beginning of the ride the vehicles are at ground level. Work has to be done to raise them up to the top of the track, and

Figure 7.5 Photo of the Pepsi Big Mac ride at Blackpool (Blackpool Pleasure Beach)

as a result they gain potential energy. The amount of work done in raising a vehicle up a vertical height h is given by:

$$\text{Work done} = \text{Force} \times \text{distance}$$
$$= m \times g \times h$$

so the **gain in potential energy or gain in PE** $= m \times g \times h$ \hfill (7.2)

The vehicles are released at the highest point of the track where they have the maximum PE. Every time they go down a slope, they lose PE and gain kinetic energy (KE). When they run up a slope, they lose KE and gain PE.

Assuming no work is done against friction, we can calculate the speed gained from rest as a vehicle falls, since the loss of PE is equal to its gain of KE. So:

$$m \times g \times h = (1/2)m \times v^2.$$

A simple pendulum is a system in which PE and KE are continually interchanging. At the limit of its swing the pendulum bob is at its highest point. Momentarily it is at rest, and so, having maximum PE, the bob has no KE. As it swings down to a central position, the pendulum bob loses PE and gains KE. At the central position the bob has

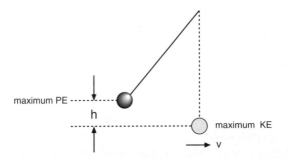

Figure 7.6 A pendulum swinging with no air resistance would continually interchange its KE and PE

maximum KE and minimum PE. And as the bob swings up to its limit on the opposite side, its KE decreases as it is converted back into PE. Then the whole process is repeated (see Fig. 7.6).

7.7 Motion in orbit under gravity

Isaac Newton thought about the behaviour of an object fired horizontally from the top of a high mountain. It would fall to sea level some way from the mountain (see Fig. 7.7); but it would go further before hitting the ground if it were fired faster from the gun. Newton thought that if it were projected at the right speed, it would go into orbit round the Earth without needing an engine to keep it going.

The direction of motion of Newton's 'satellite' would be changed by the force of gravity constantly pulling it down, but at the correct speed of projection the curvature of its path would exactly follow the curvature of the Earth. So it would be in orbit.

There are, however, two practical snags with this idea. The first is that the speed of projection works out to be over 30 000 km/h, which is much more than a cannon or even a modern gun can achieve. The second snag is that friction with the air for an object travelling this fast would probably cause it to melt.

Nevertheless the principle behind this idea is sound. Satellites are put into continuous orbit round the Earth, but they have to be lifted up at least 200 km above the

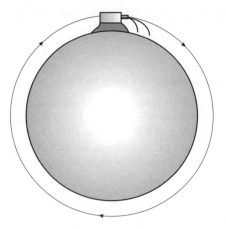

Figure 7.7 A massive gun on a high mountain might fire a missile into orbit if it were fired at the right speed

Earth's surface, giving them orbits with radii of several thousands of kilometres. At that distance from the Earth's surface, the friction of the atmosphere is not a problem as the air density is so small.

Weightlessness
Imagine that you are in a space capsule with a fellow astronaut, on top of a rocket ready to take off. If you weigh your fellow on a spring balance attached to the capsule, the balance will indicate his weight which might be 1000 N. This is a measure of the pull which the Earth exerts on him. Once the space capsule is in orbit round the Earth, it travels freely, needing no rocket to drive it. The force of gravity alone enables it to remain in orbit. No other force is needed.

This applies both to the capsule and also to everything inside it. So the astronauts also need no extra force to keep them in orbit. The spring balance weighing the astronaut will read zero, and for this reason he is said to be weightless in this situation. However gravity does still act on him.

7.8 Geostationary orbits

A satellite is put into orbit by means of a rocket or space shuttle, which must reach the correct height and speed, and launch the satellite in the correct direction. The bigger the radius of its circular orbit round the Earth, the longer it takes a satellite to go round.

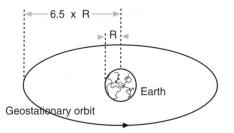

Figure 7.8 *The radius of a geostationary orbit around the Earth is about 42 000 km, or 6.5 times the Earth's radius. A geostationary satellite orbits to the east above a point on the Earth's equator*

Orbits with a radius of about 42 000 km are of special interest, because a satellite takes just 24 hours to complete one of these orbits. That means that after each rotation of the Earth the satellite will have completed one orbit and be in the same part of the sky seen from Earth. If such a satellite is **travelling due east above the equator at this height**, it will always appear to be stationary above a fixed point on the equator. As the Earth's surface rotates towards the east, the satellite will exactly keep pace with it (see Fig. 7.8). Many communications and weather satellites are in these **geostationary orbits**.

Questions

1   Why does an apple fall to the ground from the branch of the tree?
2   The Sun exerts a force on each of its planets. How do we know this?
3   Why do we think a planet exerts a force on the Sun?
4   Compare the size of the forces on the Sun and a planet, when each attracts the other.
5   If the Sun has much more mass than its planet, how will this force affect the Sun?
6   What is the weight in newtons of a mass of 75 kg on Earth where $g = 10$ N/kg?
7   The mass of a man on Earth is 80 kg. What would be (a) his mass on the Moon, and (b) his weight at the Moon's surface where $g = 1.6$ N/kg?
8   Write down the free-fall accelerations at the Moon's surface of (a) a feather, (b) a stone (Remember, the Moon has no atmosphere).
9   Estimate (a) your mass in kg, and (b) your weight in newtons. Then (c) calculate your increase in potential energy when you rise 10 m in an elevator.
10   Explain why falling objects near the Earth reach a terminal velocity.
11   Explain what is a geostationary orbit of a satellite round the Earth.

What have you learnt?

1   Do you know that masses not touching attract each other?
2   Do you know the difference between mass and weight?
3   Do you understand that different masses fall to Earth with the same accelerations?
4   Can you explain how air resistance affects a falling object?
5   Do you know how free-fall accelerations can be measured?

6 Do you know why satellites can remain in orbits around the Earth?
7 Can you explain the changes in potential energy of a mass on an escalator?
8 Do you know what affects the time needed for a satellite to orbit the Earth?
9 Do you understand the energy changes that occur when a pendulum swings?
10 Do you know why dish aerials need no directional adjustments to receive satellite data throughout the day?

8 Energy demand and usage

Objectives

After reading this chapter you should
- understand that the Sun is the major source of the Earth's energy
- know that fossil fuels are non-renewable
- know that some forms of energy are renewable
- understand that burning fossil fuels increases the carbon dioxide in the atmosphere
- understand that plants and trees absorb carbon dioxide and give out oxygen
- understand why a greenhouse is warmer on the inside than outside
- know that some gases increase the greenhouse effect on the Earth
- know that energy can be derived from the Sun's rays, tides, winds, water waves, etc.
- understand that nuclear energy can be used to generate electricity.

8.1 Introduction

A great increase in the demand for energy was seen in the nineteenth and twentieth centuries. Much of the demand was met by using **fossil fuels** like coal, natural gas and oil from which both petroleum and paraffin are refined. Fossil fuels, however, are non-renewable, and so eventually they are expected to be used up. If energy demands in the twenty first century are to be fully met, we may need to

(a) modify our lifestyle so as to reduce the demand for energy
(b) find more efficient ways of using our resources
(c) make greater use of renewable energy supplies like wind and wave energy
(d) protect the environment so that it maintains high yields of renewable energy
(e) develop a controlled fusion process to release nuclear energy from some of the hydrogen nuclei in water molecules.

8.2 Sustainable use of energy

As the world's population has continued to increase, and as all major world communities are using more and more energy, an energy crisis is developing. The world's fossil fuels of oil, natural gas and coal are being used up rapidly, and in 100 to 200 years will be gone. They are not being replaced.

Wood from trees is a renewable energy source; but in many developing communities mature trees are being cut down faster than new saplings are growing to take their place. In some regions tree felling and new planting are controlled to produce a steady and sustainable supply of wood for fuel or other uses. Without this control, wood will be in increasingly short supply. With proper control, and with a large enough area in which trees can grow, an adequate sustainable supply of wood can be produced for each community.

This will not on its own solve the world's energy crisis; but it would make it less severe. We are drawing attention here to the **increasing need for wise planning in the uses of the Earth's resources**.

8.3 Energy resources

Table 8.1 shows some of the Earth's sources of energy in renewable and non-renewable forms.

Table 8.1

Renewable sources	Non-renewable sources
Sunlight	Fossil fuels like oil, coal,
Wind energy	and natural gas
Wave energy	Nuclear energy
Tidal energy	
Energy from hydro-electric power	
Wood for fuel	

In the left column above are energies derived from the sun's energy on a daily basis, so they are regarded as renewable. On the right, the fossil fuels are stores of chemical energy. They result from the action of sunlight on plants which grew thousands of centuries ago. Nuclear energy alone does not depend on sunlight. It comes from the central parts of some atoms as described in chapters 28 and 29 on radioactivity. However, it must be clear that **the Sun is the major source of energy used on Earth** at present.

8.4 Where does energy go?

Typical fossil fuels like coal are concentrated sources of chemical energy. An electric battery is another concentrated source of energy in chemical form. **Energy in a concentrated form is easy to use**; but once it has been used to heat a building, for example, or to light a bulb, the energy is less concentrated. It becomes spread out and shared among more and more molecules. This makes it less available for further use.

Another example where some energy becomes unavailable is in a coal fired power station. The burning of the fuel produces high pressure steam in a boiler, and some heat energy which escapes. The steam rotates turbines so that useful work is done on the generators which produce electrical energy. The process by which the steam gives up its energy to the turbines is not very efficient. The steam becomes cooler, and it

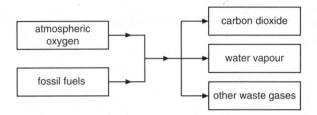

Figure 8.1 Fossil fuels burn in oxygen from the air to produce chiefly carbon dioxide and water vapour, with some other gaseous products

must be cooled further before being returned as water to the boiler; but this cooling means that much energy originally absorbed in the boiler is spread out and lost to the surroundings. Only about 25% of the original chemical energy in the coal is usefully converted into electrical energy.

The conversion of energy from one form into another is seldom 100% efficient. Energy is usually lost in machines where work has to be done against the forces of friction. This results in heat being given out to the surroundings. So global warming may result from many kinds of energy conversion.

On a large scale the universe will become warmed very slightly as centres of concentrated energy such as stars become cooler. This process will take a huge span of time.

8.5 Using fossil fuels

One of the uses of oil is as a fuel to make engines work. In a car engine the burning of oil produces heat, water vapour, carbon dioxide and some other gases. The process takes up oxygen molecules from the air, so **burning fossil fuels reduces the amount of atmospheric oxygen and increases the amount of atmospheric carbon dioxide** (see Fig. 8.1).

The balance of these gases in air is restored by plants on land and plankton in the oceans. Photosynthesis in plants involves taking in carbon dioxide from the atmosphere and energy from the Sun, while giving out oxygen. Nevertheless **the proportion of carbon dioxide in the atmosphere has been gradually increasing**, giving rise to gradual global warming.

> **Example. The car versus the runner.**
> Let's consider running a marathon, or driving a small car the same distance, i.e. about 26 miles.
>
> The car needs about two thirds of a gallon of petrol, i.e. about 3 litres of a mixture of organic molecules called alkanes, mostly octane. Octane has a chemical formula C_8H_{18}. The energy released by burning this amount of the fuel is about **110 MJ,** and the journey might take about half an hour. A runner's energy comes from food, and although food is not a fossil fuel, it does break down to produce carbon dioxide. The marathon runner needs

about 420 kJ of energy/mile in the race; so the total energy requirement of the runner is about 26×420 kJ = **11 MJ**.

Thus the car needs roughly ten times more energy than the runner, and it emits more carbon dioxide; but the journey by car is more comfortable, and it takes perhaps only one sixth of the time. The car's driver is unlikely to be exhausted by the journey. There is more about this journey in Section 11.6.

8.6 Greenhouse effect

Greenhouses are warmer inside than the air outside, because the Sun's rays pass into the greenhouse more easily than thermal radiations emitted from inside can escape (see Fig. 8.2). Glass is not good at transmitting the longer wavelength thermal radiations

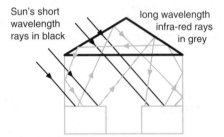

Sun's short wavelength rays in black

long wavelength infra-red rays in grey

Figure 8.2 The short wavelength rays from the Sun pass through glass into the greenhouse. Objects at room temperature emit longer wavelength infrared rays which are reflected or absorbed by the glass

emitted inside the structure. They tend to be reflected within the greenhouse or absorbed by the glass. The Sun is so hot that it radiates mainly shorter wavelength rays which pass through glass. So the **rate of receiving energy into the greenhouse can exceed its rate of escape** and whatever is inside the greenhouse warms up.

Thermal radiation is described more fully in chapter 10.

8.7 Greenhouse gases

Gases in the Earth's atmosphere which behave like glass in a greenhouse are known as greenhouse gases. **Carbon dioxide has this effect and it is one of the gases responsible for global warming**. Each day the Sun's radiation reaches parts of the Earth at a greater rate than the rate at which radiation from the Earth passes out into space. So the surface of the Earth and the air nearby is warmed. At night the net flow of radiation is out from the Earth's surface into space, and the surface cools down. However, the presence of too much atmospheric carbon dioxide causes some reflection of the outgoing radiation as well as some absorption. The atmosphere and the ground below remain warmer than they might be, as though they were in a greenhouse.

One result of gradual global warming is a reduction of the amount of ice in polar regions and a slight rise in average sea levels around the world. Low-lying coastal regions are under greater threat of flooding, so the global concentration of greenhouse gases is an important issue.

8.8 Polluting gases

Carbon dioxide is not the only gas entering the atmosphere when fossil fuels are burned. Petrol and diesel fumes from internal combustion engines are known sources of pollution, giving rise to breathing disorders like asthma in places as different as Los Angeles in California and Santiago in Chile. These areas have **heavy road traffic, low surface winds, and sunlight which reacts with motor exhaust gases converting them into more harmful products**. Nitrogen dioxide and carbon monoxide are two harmful gases which can be emitted from motor exhausts. Unless the cost of fossil fuels for vehicles becomes high enough to limit the volume of cars on the road, polluting gases may well become an increasingly serious health hazard. Industrial processes can be a second source of polluting gases, and the effluents from chemical plants need to be carefully monitored.

While some gases cause air pollution near the ground, other gases threaten the destruction of ozone gas in the upper atmosphere. Where ozone occurs at high altitude, it can absorb, and thus filter out some of the more harmful ultra-violet rays from the sun. Its destruction creates an increased threat of skin cancers in those who are not well protected from ultra-violet light.

8.9 Using solar energy

A device which uses energy from the Sun is called a solar panel. There are two main types of solar panel. They convert the Sun's radiation into either:

(a) electrical energy in the form of direct current like that from a battery of cells
(b) thermal energy in a fluid inside the panel.

The first type of panel is expensive and not very efficient, though some progress has been made with both these problems. The panels were developed to provide electrical power for transmitters in communications satellites and for other satellite uses. The amount of power developed increases with the area of the panel exposed to the direct rays of the Sun.

A typical panel for absorbing thermal energy is housed behind double glazing in a thermally insulated box. The hollow metal panel is filled with a liquid which can be pumped through a heat exchanger to warm up some water for domestic use. The amount of energy received per second in bright sunlight depends on the latitude of the site. While 1 kW/m^2 is possible in the south of France, the reception in the UK is perhaps half as much.

Solar energy is renewable in the sense that it will be continually supplied for as long as the Sun keeps supplying it, a time scale thought to be for many millions of years. However, atmospheric pollution from smoke and volcanic dust has shown that it is possible to prevent much of the Sun's radiation from reaching ground level, and there must be some concern that such pollution could spread to create many global problems.

Figure 8.3 Wind turbines on a hill near Bridgend, South Wales (Graham Bell)

8.10 Wind energy

The kinetic energy of air in the winds can be partly converted into electrical energy which is fed to the National Grid. Wind farms are sited in places regularly exposed to winds throughout the year. Fig. 8.3 shows wind turbines on a hilltop near Bridgend, South Wales. Tall pillars support the three-bladed propellers, which rotate facing into the wind. Each traces out a circle of diameter 37 m, and drives an alternator with maximum electrical output of 500 kW. The farm at Bridgend has 20 generators with an annual output of about 22 million units, or kWh, (see Section 6.5). It is enough to supply about 6000 homes.

The power available depends on both the **kinetic energy** of each cubic metre of air, and the **number of cubic metres passing each propeller per second**. The kinetic energy is **proportional to the wind velocity squared**, and the number of cubic metres per second is **proportional to the wind velocity**. So the power of the air reaching each propeller and generator is proportional to its velocity cubed. This means that if the wind speed doubled, the power reaching a propeller would increase by a factor of $2^2 \times 2 = 8$.

Wind energy is renewable, because it depends only on a continual supply of radiation from the Sun falling on the Earth's atmosphere and creating the winds.

8.11 Energy in water waves

As with wind energy, there are sites which are best suited to the conversion of water wave energy into electricity. The waves at these places are larger on average, and so

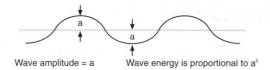

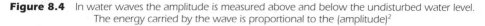

Wave amplitude = a   Wave energy is proportional to a²

Figure 8.4 In water waves the amplitude is measured above and below the undisturbed water level. The energy carried by the wave is proportional to the (amplitude)²

carry more energy than smaller waves. If we refer to the **amplitude of a wave** as the vertical height of the wave crest above the undisturbed water level (see Fig. 8.4), then we find that the energy in the wave increases in proportion to its amplitude squared. Many designs have been put forward for the energy converters, but few commercial devices are presently in use. As this form of energy is renewable, it presents attractive possibilities for the future when the cost of other forms of energy is likely to increase.

8.12 Tidal energy

The rising of tidal waters in a funnel shaped estuary twice daily represents an increase in potential energy of the water. The idea of using the change in PE of the water to generate electricity has been put into practice outside St Malo harbour in northern France. A wide concrete barrage lies across the entrance to the harbour, and incoming and outgoing water flows through openings in the barrage. As the water passes through the barrage, it drives turbines which in turn drive the electric generators. While the power output from these generators varies during a day, the system provides electrical energy at an economical rate.

It has been suggested that this system could be used in the Severn estuary in the United Kingdom, as well as in some overseas areas where the differences between high and low tides are also large. The second Severn road bridge between England and Wales was built rather than a roadway across a wide barrage in the same area, because the relative cost of the barrage, possible silting up of the estuary, and the environmental impact all favoured the bridge. While the barrage system looked possible in terms of energy conversion, it was not adopted for other practical considerations. A barrage with openings for turbines might become attractive in the future, particularly if the cost of electrical energy increases.

8.13 Hydroelectric power

Rain water drained into a reservoir high in the hills has potential energy it can give up on flowing down towards sea level. When it flows down through pipes, it exerts a pressure on turbine blades in a generator house, and the turbine drives generators to provide electric power. The amount of electrical energy this system can provide depends on the amount of local rainfall, and the vertical height of the reservoir above the generators. Advantages of this system are that it is cheaper to run than a fossil fuelled unit, and it can also be turned on and off to balance supply with demand. A coal-fired, or nuclear power plant is best operated in steady conditions. So it is possible to use a coal or nuclear unit to pump water up into a reservoir overnight when demand is low, and to use these plants and the hydroelectric unit to generate electricity when demand is high.

8.14 Using nuclear fuels

A disadvantage of using fossil fuels is that they are non-renewable. Another disadvantage relates to global warming and harmful gas emissions.

However, nuclear fission reactors do not depend on chemical reactions when they derive heat energy from sealed fuel elements. So nuclear fuels like uranium do not contribute to global warming through the production of greenhouse gases. The more developed countries have taken advantage of nuclear energy for generating electricity, but to varying degrees. Why have these countries not turned to nuclear power more wholeheartedly? The disadvantages of nuclear power stations are:

(a) the costs of construction and decommissioning after their active lives are over. If a power station has a safe working life of twenty five years, say, it must then be closed down and decommissioned. This is costly as the reactor must be taken away, and radioactive parts must be transferred to a place of safe storage for many hundreds of years
(b) the cost of uranium fuel elements
(c) the health precautions required for the plant to be operated safely
(d) the risk of environmental damage if radioactive matter were to escape. This could occur either as a result of cracks in or close to the reactor, or human error in operating the plant. The best known instance of leakage was at Chernobyl when human error was involved
(e) public concern over the effectiveness of fail-safe devices.

Despite some public concern, nuclear reactors have been in use in power stations since the 1960s, and for some countries without natural oil reserves nuclear energy has been an invaluable resource. There is more on this subject in section 29.3.

8.15 Energy for the twenty-first century

What ideas can be put forward for meeting the energy demands of future years?

(a) Power stations using nuclear fusion of heavy hydrogen nuclei may be introduced.
(b) There may be a reduction in the need for travel, reducing energy demand.
(c) Alternative renewable energy sources may be used more fully.

It may be possible to control the energy released in nuclear fusion. Like fission, fusion should release no greenhouse or polluting gases into the atmosphere. A point in favour of fusion is that its raw material is water. Just a small proportion of water molecules, about one in 6000, has the kind of hydrogen that can take part in fusion. The technology for controlling the process is not yet available, and there is no certainty that it will be. So this source of energy is by no means assured. One of its attractions, however, is that a huge amount of energy is potentially available in the world's oceans. Another possibility is that the hydrogen required for fusion may also be derived from other sources.

The world's population has been increasing and the rate of increase is growing. It seems certain that, as the population grows, the demand for energy worldwide will increase, or general standards of living will become lower. Assuming both the growth and demand increase, you may like to consider what significant energy savings could be made in your country, while essential needs of the population are met, and people have the chance to live purposeful and fulfilling lives.

1 Why are some energy resources regarded as renewable?
2 Energy cannot be destroyed, so why should there be an energy crisis?
3 (a) What might raise energy demands worldwide in the twenty-first century?
   (b) What might lower those energy demands?
4 Explain whether or not you think more nuclear energy should be used in the next hundred years.
5 (a) What advantages can you give for increasing the use of renewable energy supplies?
   (b) Which type(s) of renewable energy should be further exploited in your opinion?
6 (a) Give an example of a greenhouse gas.
   (b) What causes an increase in greenhouse gases in the atmosphere?
   (c) Explain the advantages of a reduction in global warming.
7 (a) A kilogramme of water is boiled in a kettle. List three different possible uses for the boiling water.
   (b) The same energy could be transferred to a freshwater pond if the boiling water from the kettle were poured into it. Explain why this energy could not now be used as in part (a).
8 Large areas of forest and vegetation can increase the amount of rainfall in some areas. How will the large continental land masses be affected if the forests are destroyed?
9 Give one harmful effect of pollution in (a) the atmosphere, and (b) the oceans.
10 What is the Earth's major energy supplier?

What have you learnt?

1 Do you know which forms of energy are renewable and which ones are non-renewable?
2 Can you give reasons both for and against the uses of (a) fossil fuels and (b) nuclear energy?
3 Do you know what the greenhouse effect is?
4 Do you know what causes greenhouse gases in the atmosphere?
5 Do you know what effect vegetation has on the balance of gases in the atmosphere?
6 Can you explain the dangers of global warming?
7 Can you explain why there may be a growing demand for energy in the next fifty years?
8 Do you know what makes energy unavailable for use?
9 Can you give some advantages of using renewable energy forms in the next century?
10 Do you know some causes of polluting gases in the atmosphere?
11 Do you know that the Sun is the Earth's major source of energy?
12 Do you know which atomic nuclei release energy in (a) fission, and (b) fusion?

9 Machines and energy converters

Objectives

After reading this chapter you should
- understand how a lever can change one force into another
- understand how hydraulic pressure is used for example in motor car brake systems
- know how pulleys change an effort force into one applied against a load
- know what is meant by the efficiency of a machine
- be able to explain an energy flow diagram for a heat engine
- be able to explain the energy transformations in some domestic appliances.

9.1 Introduction

We have found that energy cannot be created or destroyed. In this chapter we shall look at things which produce energy in a form we can use, and consider how well they perform their functions.

9.2 Levers

A lever is a machine and machines in general change one force into another. The force applied to a machine is called the **effort** and the machine exerts a different force against a **load**. The machine is a **force multiplier**; and the ratio of the load to the effort measures the machine's performance in scaling the effort force either up or down. For example, a lever which can lift a load weighing 12 N with an effort of only 4 N can scale up the force in the ratio 12:4. We call this ratio the mechanical advantage (MA) of the machine:

$$MA = \text{load/effort} = 12 \text{ N}/4 \text{ N} = 3 \tag{9.1}$$

A lever can be used to lift a 1 kg bag of sugar. In Fig. 9.1 different types of lever are shown. Each is a light bar pivoted at a fulcrum or hinged at a fixed point. The weight of a 1 kg bag of sugar is about 10 N which acts downwards on the lever. To support the sugar the lever must exert an equally big upward force on it. The effort applied to

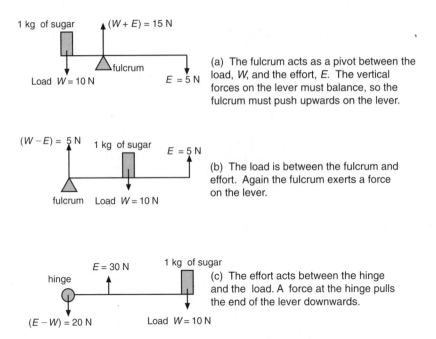

(a) The fulcrum acts as a pivot between the load, W, and the effort, E. The vertical forces on the lever must balance, so the fulcrum must push upwards on the lever.

(b) The load is between the fulcrum and effort. Again the fulcrum exerts a force on the lever.

(c) The effort acts between the hinge and the load. A force at the hinge pulls the end of the lever downwards.

Figure 9.1 *Horizontal levers are shown balanced on a pivot or hinge to support a bag of sugar*

each lever is also shown and its size indicated. To explain levers fully we apply the law of moments described in chapter 5.

___ **Self test** ___

In Fig. 9.1a, b, and c, discover the mechanical advantage of each lever.

(a. 2, b. 2, c. 1/3)

In order for the levers to be in equilibrium, the pivots or hinges must also exert forces. The total force on each lever must be zero, so the fulcrum in Fig. 9.1a must push up on the lever with a force of $(10 + 5)$ newtons. In Fig. 9.1b the required upward force is $(10 - 5)$ newtons. But in Fig. 9.1c the hinge exerts a downward force of $(30 - 10 \text{ N})$. The total upward force again equals the total downward force on the lever in each example.

Note that the forces acting at the pivot or hinge do not cause the lever to rotate. Also the clockwise and anticlockwise moments about the fulcrum of a lever in equilibrium are equal. Moments were discussed in chapter 5.

Levers are seen in nature. For example, the human arm acts as a lever in lifting a bag of sugar as in Fig. 9.2a. The elbow is like the hinge in Fig. 9.1c and the effort is applied to the forearm by means of muscles and a tendon attached to the forearm near the elbow.

Fig. 9.2b shows a nutcracker made of two hinged metal levers like the lever in Fig. 9.1b.

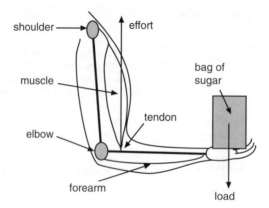

Figure 9.2(a) The effort to raise the hand is applied to the forearm by a tendon attached near the elbow

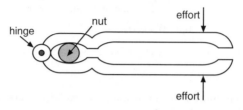

Figure 9.2(b) A nutcracker made of two levers. The forces on the nut are larger than the effort, as the nut is closer to the hinge

9.3 Hydraulics

To make oil flow along a horizontal pipeline, there must be higher pressure at one end than at the other. The oil flows towards the lower pressure, and stops when the pressure at each end of the pipe is the same. When no oil flows in a hydraulic press (see Fig. 9.3), equal pressure is exerted on two pistons in pipes of different diameter. The pressure on each piston is the same, but it exerts a proportionally larger force on the piston having the greater area. If you consider the unit area of each piston, the forces

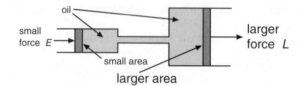

Figure 9.3 The hydraulic press enlarges a small force. Equal oil pressure acts on both pistons, and causes a larger force to act upon the larger piston

Figure 9.4 *The tipper truck on the left uses hydraulic pressure to tip its load*

are the same; but one piston has more area than the other, and so the total force on it is greater.

Force = Pressure × Area, or $F = P \times A$ (9.2)

If one piston in a hydraulic machine has five times the area of the other, it will experience five times the force.

The hydraulic press may be used in a coal mine to support the roof, or in a garage to lift a car high enough for work to be done underneath it. Machines like farm tractors and excavators operate with oil pressures greater than 150 times atmospheric pressure. Hydraulic machines which use strong high-pressure tubing can create surprisingly large forces. On the left of Fig. 9.4 is a tipper truck capable of carrying 150 tonnes of ore (i.e. 150 000 kg), excavated in an opencast South African mine. It can tip out its load using pistons driven by the oil under pressure from a reservoir in its engine. The vehicle has very large tyres to spread its weight over enough area of the ground to prevent the pressure making it sink in up to the axles. The truck is too large to be allowed onto public roads. It had to be assembled on site in the mine from a number of parts which were brought in separately. Notice the cars in the mine which carry flags on tall masts to reduce their chance of being run over by the tipper trucks. The truck drivers can see the flags and so avoid the cars (Notice also the size of the excavator's shovel at the foot of the cliff. It is operated by means of levers, pulleys and cables).

An advantage of using a liquid in the hydraulic press is that liquids retain the same density and are not compressible. They are not 'springy' like gases, and this enables a skilled machine operator to direct the very large force of the machine with great precision by means of just a few levers. These operate valves controlling the flow of oil in the pressure tubing and cylinders of the machine.

9.4 Car brakes

Many disc braking systems use both levers and hydraulic presses to force pressure pads onto the rotating discs. Fig. 9.5 shows how the foot pedal of the vehicle acts as a lever to enlarge the effort applied to it. But notice that the brake fluid applies an enlarged force to the pressure pads between which the discs rotate, because the area of each piston at B is greater than the area of the piston at A. The brake fluid is oil which transmits the pressure from the master cylinder to each piston acting on the brake pads.

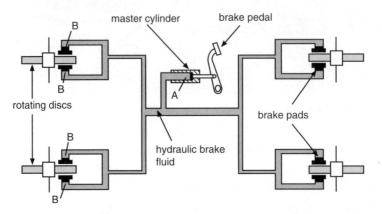

Figure 9.5 A hydraulic brake system with four disc brakes. The oil transfers the pressure from the master cylinder to the brake pads which grip the rotating discs and slow them down by friction. Most of the energy of the moving vehicle is converted into heat in the discs and pads, and carried away by the air

9.5 Pulleys

A pulley system is a machine for raising a load by means of a smaller effort. Fig. 9.6 shows a load suspended on a moveable lower pulley block. The tension produced by the effort force is transferred by the rope around the pulleys, so that the total upward force on the lower pulley block is enlarged. In the example shown, the rope pulls upwards on the lower block with a total force four times bigger than the effort. This force must balance the load plus the weight of the lower block so the load can be larger than the effort.

Machines cannot do more work than the energy that is put into them. So if a heavy load is raised using a smaller effort, the load will not move as far as the effort. Remember:

work = force × distance moved   (9.3)

The load × distance load moves is usually less than the effort × distance effort moves, because of friction and the weight of the lower pulley block.

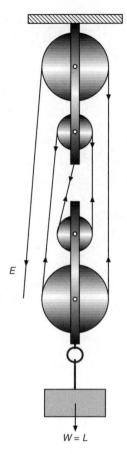

E

$W = L$

Figure 9.6 The tension in the rope acts upwards on the lower pulley block at four places, so increasing the load that can be lifted

9.6 Gears

Effort forces are exerted on the pedals of a bicycle (see Fig. 9.7) to drive it along. Through the chain and gear wheels these forces enable the cycle and rider to cross level ground against the forces of friction.

Suppose the chain is linked to the pedal wheel of 42 teeth and to the rear cog wheel of 12 teeth. Then one revolution of the pedals makes the chain move by 42 links. So the back wheel makes 42/12 revolutions, i.e. 3.5 revolutions. Using this gearing the cyclist can exert a big enough force to ride at speed on level ground.

Now suppose the cyclist changes to nearly the lowest possible gearing, with 24 teeth on the pedal wheel and 24 teeth on the rear wheel. One rotation of the pedals results in one revolution of the rear wheel.

In the higher gear, one rotation of the pedals needed 3.5 times as much work as in the lower gear. To travel at a certain speed on horizontal ground, the rider must exert a larger force using the higher gear. At a slower speed and in the lower gear, the rider could apply a much smaller force on the pedals; but it would be possible to drive the

Figure 9.7 A mountain bicycle has low gears for uphill climbs, and knobbly tyre treads to maintain good grip on rough ground
(Falcon Cycles Ltd)

cycle up a slope by using the larger pedal force and doing work to increase its potential energy as well as driving it along. So we use low gears in order to cycle uphill, but travel more slowly than on level ground.

9.7 Efficiency

The efficiency of a machine is:

efficiency = (useful work output/total work input) × 100%   (9.4)

Machines are always less than 100% efficient. One reason for this is that some work is done against friction forces when the surfaces of moving parts of a machine slide over each other, as for example, when cog wheels interlock. A second reason is that work is done against gravity in lifting parts of a machine such as the lower pulley block as a load is raised using pulleys. If more work is done either against friction, or in raising parts of a machine, then the useful work output and the efficiency of the machine will both be reduced. Remember **useful work is work done on the load**.

9.8 Inclined plane

A pulley at the top of an inclined plane enables a load to be lifted more easily than without this device. For example a garden roller can be hauled up an incline to the back of vehicle using a smaller force than its weight. The same work is done to increase

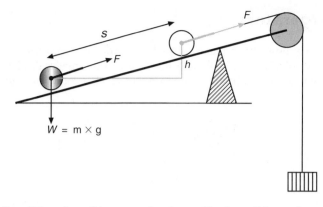

Figure 9.8 To pull the roller a distance s up the slope, with a force F, the work needed is $F \times s$. The gain in PE of the roller is m × g × h. This machine allows a smaller force to move further in doing useful work

its potential energy, but this is done with a smaller force operating for a greater distance along the incline. Fig. 9.8 shows that the roller moves vertically a distance h, when the effort moves a larger distance s. The efficiency of this system is:

$$\text{efficiency} = (m \times g \times h/F \times s) \times 100\% \tag{9.5}$$

since $m \times g \times h$ is the useful gain in PE when the effort F moves a distance, s, up the incline.

9.9 Heat engines

Heat flowing from a high temperature to a lower one can be made to do some work. Q joules of heat at high temperature in a cylinder of an internal combustion engine can cause gases to expand and do W joules of work. The remaining energy flows out of the

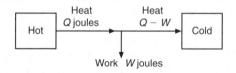

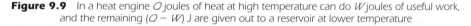

Figure 9.9 In a heat engine Q joules of heat at high temperature can do W joules of useful work, and the remaining $(Q - W)$ J are given out to a reservoir at lower temperature

engine at lower temperature, mainly in the exhaust pipe of the engine. Fig. 9.9 illustrates the flow of energy.

The efficiency of the heat engine is:

$$\text{efficiency} = (\text{useful work output/energy input}) \times 100\% = (W/Q) \times 100\% \tag{9.6}$$

9.10 Refrigerators

Heat spontaneously flows from places at higher temperature towards lower temperatures. So in a mixture of hot and cold substances the hotter materials give out heat and the colder ones take it in. Eventually the substances reach a common temperature.

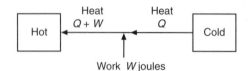

Figure 9.10   In a refrigerator heat flows from food at lower temperature into air which is warmer. W joules of work are needed to reverse the natural heat flow direction

In a refrigerator the reverse process is made to happen, so that cooler materials are made colder still and the warmer surroundings become heated. The process requires an energy input represented by the symbol W in the energy flow chart in Fig. 9.10. Chapter 11 includes more about refrigerators.

9.11 Heat pumps

The refrigerator can be used as a heat pump when the heat output at the higher temperature is used to warm something up. An air conditioning plant is a typical example. The heat pump's performance or its 'coefficient of performance' (C.O.P) is:

$$\text{C.O.P.} = (\text{Energy delivered at high temperature/total energy input}) \tag{9.7}$$

C.O.P. values of 3.5 can be achieved in practice, which means that 3.5 kWh of energy can be used to heat a building when only 1.0 kWh of electrical energy is supplied to the pumping system.

This principle was used in 1951 for the heating system of the Royal Festival Hall which stands on the south bank of the river Thames in London. Heat was pumped from the cool river into the warmer Festival Hall leaving the river even cooler. The system needed less energy input than the heat delivered into the building. Yet the law of conservation of energy still applies. In the above example at least 2.5 kWh of energy extracted from the cold source, i.e. the river, plus the energy output from the pump, provide the total energy of 3.5 kWh for heating the building, when only 1.0 kWh of energy was supplied to the pump.

The advantage of a heat pump is that it is cheap to run, but a disadvantage is that it is costly to buy initially and to install.

9.12 Electrical energy in the home

Any electrical appliance gives out some energy to its surroundings. This is unavoidable, and it causes loss of efficiency in each instance. For example:

(a) A toaster. Electrical energy from the mains supply is converted into heat when the

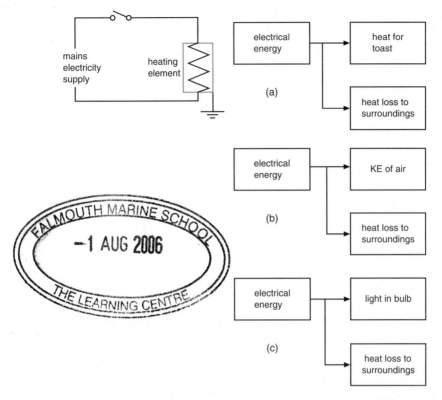

Figure 9.11 This figure shows energy changes when electrical energy is used (a) in a toaster, (b) in a fan and (c) in a filament light bulb. In each example some energy flows into the surroundings

circuit is completed by a heating element in the toaster. Some of the heat browns the toast, while some of it escapes and spreads out into the surrounding air.

(b) A cooling fan. Electrical energy from the mains supply is converted into the fan's rotation by an a.c. motor. The fan causes a steady flow of air, which cools the skin by speeding the rate of evaporation of water or sweat from its surface. But the kinetic energy of the air is shared by more and more molecules, which results in a slight warming effect in the surroundings.

(c) A filament lamp. The filament of a bulb completing an electric circuit becomes white hot when a current flows through it. The light emitted is radiated away from the bulb, which also becomes a heat source and warms the surroundings. Fig. 9.11 shows these energy changes.

Questions

1 A machine with a mechanical advantage of 4 raises a load of 20 N. What effort is applied to it?
2 If the load of 20 N is raised 3 m when an effort of 8 N moves 10 m,
   (a) How much work is done on the load?
   (b) How much work is done by the effort?
   (c) What is the efficiency of the system?

3 (a) Make a diagram of a hydraulic press with pistons having areas in the ratio 3:1.
  (b) Why is it filled with oil instead of air?
  (c) Which piston has the greater force on it?
  (d) If the effort on the smaller piston is 10 N what load could the press support?

4 A piston in a hydraulic machine is required to exert a force of 20 000 N. The extra pressure in the oil inside the machine is 2.5 MPa. What must be the area of the piston?

5 Where is friction likely to occur in a pulley system?

6 (a) How would you decide whether to install a heat pump to heat a large building?
  (b) What change would make the pump more cost effective?

7 (a) Why are machines never 100% efficient?
  (b) Where does the energy go if it does not produce useful work?
  (c) What could you do to try to improve the efficiency of a machine?

___ **What have you learnt?** ___

1 Do you understand that machines change one force into another?
2 Do you know what the 'mechanical advantage' of a machine is?
3 Can you explain the meaning of efficiency of a machine?
4 Can you explain how a hydraulic press works?
5 Do you know the way a car's brakes work?
6 Do you understand what a heat engine does?
7 Do you know that energy used in domestic appliances becomes dispersed?
8 Can you explain why refrigerators need energy to keep food cool?
9 Could you calculate the efficiency of a machine or heat engine?
10 Can you apply the principle of moments to a pivoted lever?

10 Transfer of heat energy

10.1 Introduction

Keeping warm is a basic human need. In winter we wear 'warm' clothing to feel comfortable and we work indoors in heated rooms. This chapter deals with the principles of home heating, heat flow and temperature control.

10.2 Conduction

Ducted warm air was first introduced by the Romans to heat their houses. Stone floors were built on stacks of tiles allowing enough space for warm air from a fire to circulate beneath the floor. In order to warm up the rooms, heat had to be conducted upwards through the stone floor (Fig. 10.1).

Heat **conduction** also takes place in modern metal central heating radiators filled with hot water.

The direction of heat flow is from places at higher temperatures towards places at lower temperatures. So heat is conducted towards the outside of the radiator when the water it contains is hotter than the surrounding air.

Similarly the handle of a poker gets hot if its other end is placed in a fire. Again the heat flows towards the cooler parts of the metal by conduction.

The kinetic theory of matter explains conduction as heat transfer through a solid whose particles continually vibrate. **The particles vibrate more violently as the**

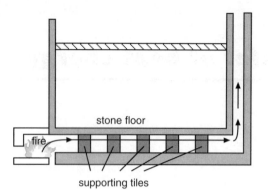

Figure 10.1 *Under-floor heating warms the room as the stone floor conducts the heat upwards*

temperature is raised. For example, the end of a poker in the fire has atoms which vibrate more violently than those further away from the fire. However, the more violently vibrating atoms affect their neighbouring atoms, which begin to vibrate more, and so that part of the rod begins to heat up. Gradually the atomic vibrations are transferred along to the rest of the rod which heats up too (see Fig. 10.2).

While conduction takes place in any substance which has parts at different temperatures, some materials are better conductors than others. For example, if a wooden rod is used instead of a metal poker, the end outside the fire stays cool, while the end in the fire gets hot enough to burn. By comparison with metals, wood is a poor conductor of heat.

Metals are particularly good heat conductors because they contain free electrons as well as vibrating molecules. Where the metal is hot, the electrons on average have more kinetic energy, which they transfer through the solid to the cooler parts.

Air is a poor heat conductor. On a cold day pockets of air in a string vest, trapped against the body by a shirt, can reduce the rate of heat loss and act as a good thermal insulator. Bad conductors are by definition good insulators, because they limit the rate of heat flow. In the attic of a house, fibreglass loft insulation traps pockets of air and reduces heat flow into the attic from the warm rooms below. In nature, wool or fur provides thermal insulation for animals in cold climates by trapping a layer of air. The teddy bears which children enjoy have woolly coats which feel warm to touch, because they do not conduct away a child's body heat.

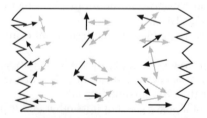

Figure 10.2 *A metal has vibrating molecules (grey arrows), and free electrons moving with KE (black arrows). The arrows are longer at the right where the metal is hotter, and particles move with more energy. The particles disturb each other, and energy is conducted to the particles on the left*

10.3 Convection

If you hold your face slightly above a hot central heating radiator, you can feel hot air rising past your face. Next to the hot surface of the radiator, the air is warmed up. Molecules of warm air move about faster and spread out more, so that the air becomes less dense than its surroundings. It rises up, taking energy from the radiator and colder, denser air sinks down to take its place.

The rising of warmer, less dense air and the simultaneous falling of cooler denser air is called **convection.** The process can occur in a liquid as well as in a gas. For example, a saucepan or electric kettle filled with water is always heated at the bottom. The heated water rises and cooler water sinks to be heated in its turn. So eventually the whole of the water is heated, up to boiling point if required (see Fig. 10.3).

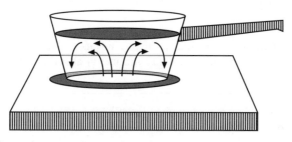

Figure 10.3 Convection occurs in water heated on a stove. In the centre the warmer water rises while denser, cooler water flows down at the edges

On a large scale we often see the results of convection currents in the air in summer, when warm moist air rises and cumulus clouds form. On a smaller scale you have only to notice that the ice box inside a refrigerator is at the top, so that the air will circulate by convection. Colder, denser air sinks down to cool the lower parts as warmer, less dense air rises to take its place and become cooled in its turn.

10.4 Radiation

If you are near a glowing electric bar fire (see Fig. 10.4), you may feel yourself getting hot. Yet the air between you and the fire is a poor heat conductor. You can feel the heating effect even at floor level so the energy transfer cannot be by convection. This third way of transferring heat energy is by radiation from the fire by electromagnetic waves, most of which have wavelengths slightly longer than visible light. It is called **infra-red or thermal radiation,** and it forms part of the **electromagnetic spectrum**.

Though its wavelengths are longer than those of light waves, infra-red radiation can transfer energy, even through an empty space, at the speed of light. Just as light reaches us from the Sun, infra-red rays also bring us energy from the Sun, which emits a continuous band of radiations of different wavelengths. Life on Earth has evolved and been sustained by the energy received from the Sun. So this is, for humans, a most important method of energy transfer.

In experiments in the laboratory we find that dull black surfaces emit more

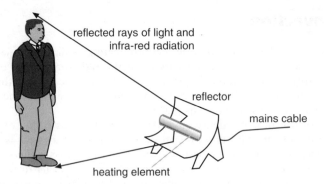

Figure 10.4 The heating element of a one bar electric fire emits both light and infrared radiation. The reflector is a shiny curved piece of metal, which reflects the rays while it remains only slightly warm. A wire safety grill (not shown) prevents large objects from falling onto the hot element

thermal radiation than shiny metallic or white surfaces. And we also find that dull black surfaces are better than shiny or white surfaces at absorbing thermal radiation.

Self test

Why do many tennis players wear white clothes?

10.5 Heat transfer

There are therefore **three methods of heat transfer** from place to place: **conduction, convection,** and **radiation**. The natural flow of heat by any one of these methods is from higher towards lower temperatures. So heat energy tends to spread out from concentrations at high temperature; and as energy spreads out, it becomes more and more difficult to use (see section 8.4).

10.6 Designing a house

Around the world the designs of houses reflect the need to achieve comfortable air temperatures for those who are inside. In hot climates there are shaded entrances, openings to allow air to circulate, and sometimes shady inner courtyards with fountains or pools where heat from the air is absorbed as water evaporates. These houses have white outside walls, because white surfaces tend to reflect the direct rays of the Sun rather than to absorb them and become warmer. The greenhouse effect is reduced by putting the larger windows in walls facing away from direct sunlight.

10.7 Keeping a house warm in winter

In Britain, where the climate is not so warm, house architects face different problems. The main problem is to ensure that houses will be adequately heated during the winter.

Radiators filled with hot water from a central boiler are often used in Britain. The radiator is poorly named, since it gives off about 80% of its heat energy by convection and only about 20% by radiation. The rate of giving out heat is directly proportional to the outer surface area of the radiator; this is sometimes increased by having extra metal sheets behind the water-filled box but attached to the main structure.

In a heated room the temperature should remain constant at an acceptable level. So the energy supplied by the radiator(s) must just balance the heat lost through the walls, windows, ceiling and floor.

The rate of heat flow through a given material is proportional to the difference in temperatures on each side of the material. Different materials, however, transfer heat at different rates. Some typical rates are shown in Table 10.1, where they are expressed

Table 10.1 U-values (Watts/m²/°C)

Single glazing	metal frame	5.6
	wood frame	4.3
Double glazing	metal frame	3.2
	wood frame	2.5
Solid 220 mm wall with 16 mm plaster		2.1
Cavity wall filled with insulation		0.5
Uninsulated roof		2.0
Insulated roof		0.3
Solid floor without carpets		1.0

as U-values. The rates are in watts per square metre of surface area, when the temperature differs on each side of the material by 1°C.

To calculate the rate of heat loss in watts through the windows of a room, say, we multiply the U-value for windows × surface area of windows × difference between inside and outside temperatures.

In symbols this would be:

$$\text{rate of heat loss} = U \times A \times (T_2 - T_1) \text{ watts} \tag{10.1}$$

To find the total rate of heat loss from the room, we need to work out the rates for each part of the room, and add them all up.

There is a considerable reduction in U-value when single glazed windows are changed to double glazing, and also when an uninsulated roof is changed to an insulated one. Similarly, there is a reduction in U-value for walls, when a solid external wall is replaced with one having an air cavity between inside and outside layers. These are typical of the energy-saving features which are built into modern UK houses.

> **Example**
> Imagine a bungalow with a total floor area of 100 m² and the same ceiling area. It has a total wall area of 150 m², a total area of windows and doors of 30 m². The windows and doors are double glazed with metal frames, the cavity walls are filled with insulating material, the solid floors are

carpeted, and the roof is insulated. We wish to calculate the rate of heat loss from this bungalow if the whole of the inside is kept at 15°C when the outside temperature is 5°C.

Answer

Each rate of heat loss = U-value × area in m² × temperature difference. So, for the different parts, the rates are given by:

Walls	$R = 0.5 \times 150 \times 10$	= 750 W
Windows and doors	$R = 3.2 \times 30 \times 10$	= 960 W
Floors	$R = 0.3 \times 100 \times 10$	= 300 W
Ceilings	$R = 0.3 \times 100 \times 10$	= 300 W
	Total rate of heat loss	= 2310 W

Self test

If the roof of the bungalow had been uninsulated, but everything else remained the same, show that the rate of heat loss would have risen to 4010 W.

10.8 Reducing heatflow

The figures in Table 10.1 and the worked example above illustrate an important principle. The rate of heat loss through, say, a ceiling, depends on whether the ceiling is insulated. If the ceiling is insulated by a layer of felt or some similar material covering it, the rate of heat loss is much reduced.

Similarly, if water pipes and the water tank in the roof of a house are wrapped in felt lagging, the water loses heat at a slower rate on cold nights, so it has less chance of freezing. While metals conduct heat energy readily, others materials like wood and felt are poor at transferring heat.

Lagging a hot body, e.g. a pipe containing hot water, may considerably reduce the rate of heat loss from the water. The slow rate of heat flow through the lagging limits the rate at which heat can be transferred from the hot water inside the pipe to the surrounding air.

A thermos or vacuum flask is for keeping a liquid either warmer or cooler than the surrounding air. It does this by reducing the rate at which heat is transferred to or from the surroundings. Fig. 10.5 shows the vacuum flask made of two flasks, one sealed inside the other, with an evacuated space between the flasks. The glass surface between the flasks is silvered like a mirror, so that it reflects infra-red radiation, preventing it from passing into or out of the flask assembly. Conduction is small through glass and air, and as there is very little air between the glass surfaces, convection effects are small. A bung of rubber or cork seals the opening at the top of the flask preventing convection losses from hot liquids.

Self test

There is no lid on the top-loading frozen food chest in the supermarket. Why do cold foods not warm up by convection in the air? (Hint: where is the air most dense?)

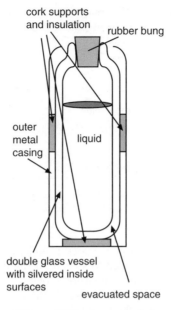

Figure 10.5 A vacuum flask

10.9 Control of heat output

In a central heating system less heat is needed when we sleep, or when the days are warmer. The output from the system is controlled by a temperature sensor in the house. Hot water passes through a closed loop of pipes and radiators before it returns via a pump to the heater (see Fig. 10.6). The temperature sensor provides the electronic data used to switch the pump on and off. In addition to the temperature sensor, there is also a **time switch** which limits the times of day when the system is in use. Each of the system controls is an electronic switch, which is either on or off.

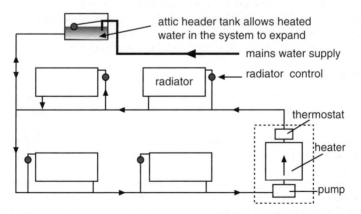

Figure 10.6 A central heating system. Hot water is circulated by means of a pump. Each radiator is controlled separately, being filled by hotter, less dense water at the top while cooler, denser water leaves at the bottom

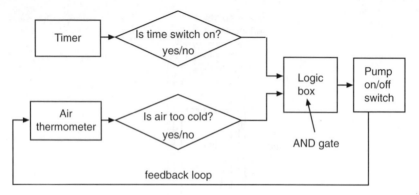

Figure 10.7 *The logic diagram for a central heating system includes a feedback loop. The pump is turned on when both inputs to the logic box are on, meaning 'yes'*

It would be wrong to call the water heater a 'boiler', as it does not boil the water. Another sensor in the hot water is used to switch off the heater before the water boils. Therefore, the water pumped to the radiators is below boiling point, and bubbles of steam are not created in the circuit. The sensor in the house measures the air temperature and turns off the pump when the temperature reaches a preset value. A block diagram showing the logic of the controls for the pump is shown in Fig. 10.7. A similar logic diagram for the heater is not shown. It includes a water thermometer in place of the air thermometer in Fig. 10.7, so that the boiler contains hot water when the timeswitch is on. This means that there is hot water available when the pump is turned on.

Fig. 10.7 includes an example of a **feedback loop,** which gives automatic control of a process. Many examples of feedback loops occur in electronic controls. Here sensors control the pump and the heater which effect the air and water temperatures, which affect the sensors which control the pump and heater, which affects the water ... you see why it is called a feedback loop? Feedback loops are widely used in automatic processes like turning on the street lights when it gets dark, or keeping the right temperature and humidity in greenhouses, or flying an aircraft on automatic pilot, or assembling motor cars using robots for welding.

The logic box which controls the pump in Fig. 10.7 has two inputs and one output. The output will only turn on the pump when both of its input lines are turned on. That means the time switch must be on, **and** the air in the house must be colder than a preset value. Otherwise the pump will be turned off. A logic box which behaves like this is called an **and gate**. We shall return to logic gates in chapter 25.

Questions

1 The speed of light in space is about 3.0×10^8 m/s. How long will it take thermal radiation to reach the Earth from the Sun which is about 1.5×10^{11} m away?

2 How does heat flow by conduction in a solid which is non-metallic?

3 Why are metals particularly good heat conductors as well as electrical conductors?

4 Refrigerators often have the cold box inside at the top. How does this help to cool all the space inside?

5 A string vest is mainly holes. How can it keep your body warm beneath a shirt on a cold day?

6   Do you think white polar bears or black polar bears would lose more body heat during the perpetual darkness of the winter months near the poles? Give a reason.
7   Look at Table 10.1. From the information supplied, design a small house, and estimate its rate of heat loss when the surrounding air and ground is 10°C cooler than the air inside the house.
8   How does a thermos flask prevent hot liquids from cooling rapidly?
9   Explain how a lighting system can be controlled by using a feedback loop.
10 Design a solar panel for absorbing energy from the Sun's rays and storing it in a hot water tank.
11 Name two good thermal insulators, and give a typical use for one of them.

What have you learnt?

1   Do you know that molecular vibrations are important for heat conduction in solids?
2   Can you explain the convection of heat in liquids and gases?
3   Do you know how free electrons increase the heat conductivity in metals?
4   Do you know how to use U-values to find the rate of heat flow through walls and windows?
5   Do you know why lagging metal pipes reduces their rate of heat exchange with their surroundings?
6   Do you know how a central heating system works?
7   Can you explain the advantages of keeping a liquid hot or cold in a thermos flask?
8   Do you know what a feedback loop is?
9   Do you know what makes an object good at radiating or absorbing thermal radiation?

11 | Changes in heat content

Objectives

After completing this chapter you should
- know how temperatures are measured
- understand what is meant by the thermal capacity of an object, and the specific heat capacity of a substance
- be able to explain the equation relating an object's gain in internal energy to its increase in temperature
- know how to measure the specific heat capacity of a substance
- know that water has a high specific heat capacity
- know that latent heat is involved in changes of state without changes of temperature.

11.1 Measuring temperature

Thermometers are temperature sensors. They use the physical properties of matter which change as a substance is warmed up. For example mercury is a liquid which expands more than glass when they warm up. A bulb of mercury is sealed as in Fig. 11.1 to a narrow glass capillary tube. It is used to measure temperatures between $-10°C$ and $110°C$, as the top of the mercury column moves along the glass tube when the temperature alters within this range. Another type of thermometer uses the change in electrical conduction of some materials as they are warmed up.

Temperature readings are related to a scale of temperatures based on fixed points. The lower fixed point is the temperature of pure melting ice (0°C on the Celsius scale or 273K). The upper fixed point, 100°C or 373K, is the temperature of steam over boiling water at a pressure of one atmosphere, or 76 cm of mercury.

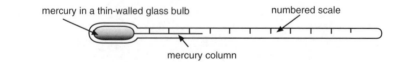

mercury in a thin-walled glass bulb numbered scale

mercury column

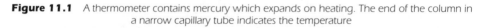

Figure 11.1 A thermometer contains mercury which expands on heating. The end of the column in a narrow capillary tube indicates the temperature

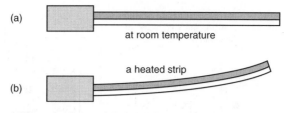

(a)

at room temperature

a heated strip

(b)

Figure 11.2 A cold bimetal strip is straight in (a), but it curls up when heated as in (b). The lower metal expands more when heated, and so it is on the outside of the curve

Bimetal strips indicate their temperatures by their shapes. The two metals are bonded to form a single thin strip. When the strip is heated, one metal expands more than the other, and the strip curves as a result (see Fig. 11.2). When cooled, it straightens to its original shape. So by fixing one end of the strip, one may use the position of the other end to indicate various temperatures.

11.2 Heating things up

When an object loses heat energy, it cools down; its temperature falls. Conversely, when energy is supplied to an object, its temperature rises as internal energy is stored in its molecules. **The amount of energy needed to cause an object's temperature rise of 1°C or 1 K, is called the thermal capacity** of the object. Thermal capacities are measured in joules per kelvin (J/K). In general:

heat energy gained = thermal capacity × temperature rise   (11.1)

Thermal capacity depends on the mass of an object and on the material it is made from. Each kilogram of a substance needs a certain amount of energy to cause a temperature rise of 1°C, or 1 K. This quantity is called the **specific heat capacity of the substance**. So we may write:

thermal capacity = mass × specific heat capacity   (11.2)

The general expression for temperature rises is given by an equation which takes into account the mass of material, what it is made of, and the amount of energy being absorbed. In general, when heat is absorbed:-

heat energy gain = mass × its specific heat capacity × temperature rise   (11.3)

Each substance has its own specific heat capacity, c. In symbols the equation is:

$$Q = m \times c \times (T_2 - T_1)$$ (11.4)

where Q is in joules, m in kg, T in kelvins, and c in J/kg K. T_2 and T_1 are the final and starting temperatures respectively. Table 11.1 shows values of c for different materials in joules per kilogram degree K, (J/kg K).

Table 11.1 Specific heat capacities of some substances

Substance	Value (J/kg K)	Substances	Value (J/kg K)
Water	4200	Copper	390
Meths	2500	Lead	130
Ice	2100	PVC	about 1000

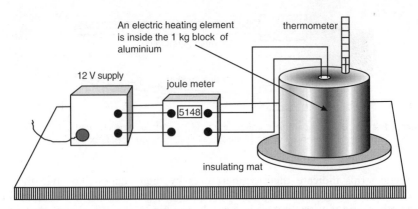

An electric heating element is inside the 1 kg block of aluminium

12 V supply

joule meter

5148

thermometer

insulating mat

Figure 11.3 Apparatus for finding the specific heat capacity of aluminium. The thermometer bulb is immersed in alcohol to maintain good thermal contact with the block.

11.3 Measuring specific heat capacity of aluminium

A convenient method for measuring specific heat capacity is to supply the energy electrically (see Fig. 11.3). The amount of energy supplied to the immersion heater is registered on the joulemeter, and the temperature rise it produces is indicated on the thermometer.

Typical readings are as follows:

Mass of aluminium block	1.0 kg
Initial temperature	20.0°C
Initial joulemeter reading	12 200 J
Final joulemeter reading	16 650 J
Final temperature	25.0°C

From these readings it follows that:

1.0 kg of aluminium is heated through 5.0°C, (or 5.0 K),
   by $(16\ 650 - 12\ 200)$ J = 4450 J.
So 1.0 kg of aluminium is heated through 1.0 K by (4450/5.0) J = 890 J.
The specific heat capacity of aluminium = 890 J/kg K.

Self test

500 g of a liquid requires 6000 J to raise its temperature from 20°C to 23°C. Show that the specific heat capacity of the liquid is 4000 J/kg K.

11.4 The high specific heat capacity of water

Table 11.1 shows that water has a particularly large specific heat capacity. Therefore, a large amount of energy has to be taken in or given out for a comparatively small change in water temperature.

This affects the climate on land close to the sea or large lakes. The specific heat capacity of water is about 5 times greater than that of earth or sand. This means that, in spring, the land temperature near large masses of water does not rise as quickly as the temperature inland. In the autumn the temperature inland falls much faster than the temperature near the coast. The presence of large masses of water tends to slow down both the cooling and warming of the land nearby. The mean daily temperatures in London, close to the sea, vary by about 11°C over the year. At the same latitude as London is Irkutsk, Siberia where the mean daily temperatures vary by 50°C over a year. Irkutsk is in the middle of the large Asian land mass.

Because of the high specific heat capacity of water, a good deal of energy is needed to boil a kettle. Once water is hot, however, it has plenty of energy to give out. So water is a suitable liquid to carry heat to radiators in a central heating system from which it will heat rooms. If the specific heat capacity of water were 10 times smaller, water would be a much less convenient carrier of energy, because so much more water would have to be pumped for the system to deliver the same amount of heat to places being warmed.

11.5 Energy to keep a house warm

In order to warm up a house, the heating system must heat both the air it contains and the material of the walls, floors and ceilings. It takes much more energy to warm up the walls than the air in the room. This means that the walls can act as a reservoir of heat from which the air can be warmed up many times. So the air can be changed and kept fresh, without the house cooling down too much. An open fire which loses much of its heat to the surroundings via the chimney can also warm up the chimney bricks and walls of the house. It may take time for the reservoir of heat to be established, and a chimney built into an internal wall of a house may be more effective than one in an external wall, because more of its energy output flows into the heat reservoir.

Once the walls and air of a house have been warmed to an acceptable temperature, the heating system is used to maintain the established conditions as described in the previous chapter. This means that heat must be supplied at the same rate that it is lost to the surroundings. This is an example of **dynamic equilibrium**. It is rather like filling a bucket with a hole in the bottom with water from a tap. Provided water flows into the bucket at the same rate that it leaks out, the level of water in the bucket remains constant.

11.6 Heating and changes of state

The temperature of a mixture of ice cubes and water stirred in a beaker is 0°C. At first it remains at this temperature when the beaker is gently heated by means of a bunsen burner. Only when all the ice has melted does further heating result in a rise in temperature. So some heat was supplied to the beaker when the ice was melting, and yet the temperature did not rise.

Further heating of the beaker of water causes the temperature to rise as far as 100°C. Then the temperature remains steady again whilst further heat is added. The water is now boiling, and it changes its state into water vapour. Because heat was needed to melt the ice and to boil the water, we can say that both these changes needed an increase in internal energy. This energy is needed to break the bonds

Table 11.2 Specific latent heats of some substances (J/kg)

	Fusion		Vaporisation
Lead	21 000	Ether	370 000
Copper	181 000	Ethanol	840 000
Ice	336 000	Water	2 260 000

between molecules whenever a solid melts. When a liquid boils, the extra energy is needed to increase the average distance between the molecules.

We call the extra energy needed to change the state of matter from solid to liquid its **latent heat of fusion.** The extra energy needed to change the state of matter from liquid to vapour is called its **latent heat of vaporization.** Specific latent heats are the energies needed to change the state of 1 kg of a substance without any change of temperature. Specific latent heats of fusion and vaporization are measured in joules per kilogram (see Table 11.2).

On a hot day elephants like to be hosed down with water. As the water evaporates, it absorbs energy from the elephants' hides. The blood vessels in their ears are close to the surface so cooling of the ears helps to cool the elephants' blood.

Humans also need to keep their body temperature nearly constant and close to 37°C. During violent activity, you get hot and your body temperature rises. Then you perspire from sweat glands near the skin's surface and, as the sweat evaporates, it absorbs latent heat from your body. By this means your body gives up heat to the evaporating sweat, and your temperature falls. So even without hosing down your body, there is still a mechanism with which you stabilise your own temperature.

In chapter 8 we said that a marathon runner needs about 11 MJ of energy for the race. Most of this energy becomes heat which must be released from the body, so as to maintain its steady temperature.

The body is mainly water, so we shall assume it behaves like 70 kg of water. Water has a specific heat capacity of $c = 4200$ J/kg.K. How hot would the runner become if the heat were not released into the air? From equation 11.4:

$$11 \times 10^6 \text{ J} = 70 \text{ kg} \times 4200 \text{ J/kg.K} \times \text{temperature rise}$$

temperature rise $= 11 \times 10^6 / 70 \times 4200 = 37.4$ K, or 37.4°C, giving a body temperature of about 74°C.

This is fatally high and no athlete could survive at such a temperature. Some of the energy will be released as the athlete breathes out gases warmed by the body, but most of the heat loss comes from the evaporation of body fluid. On a hot day this loss of fluid may be as much as 3 kg, and the runner plans for this by drinking perhaps 1 kg of water before the race and another 1 kg during the race. The remaining fluid is released from carbohydrate foods in the body as they are chemically broken down.

To evaporate 1 kg of water requires about 2.3 MJ. This means that the body can release 3 kg × 2.3 MJ/kg = 6.9 MJ of heat, which is well on the way to solving the problem. As the runner moves through the air, there will be a further loss of energy by conduction and convection to the surroundings. So, providing the runner has an adequate water supply, it is possible for his body to remain at an acceptable temperature, i.e. below about 41°C.

Questions

1 A 50 W heater warms an object from 20°C to 35°C in 300 seconds. What is the thermal capacity of the object?
2 In practice the thermal capacity is less than that shown in the calculation of question 1. What was assumed that may not have been accurate?
3 How much heat is needed to warm up 2 kg of water from 20°C to 35°C? (See Table 11.1)
4 A heated pure solid maintains the same temperature during melting. What happens to the energy it absorbs?
5 Why does evaporation cause cooling?
6 A joulemeter shows 3300 J were delivered to a mass of 2 kg when its temperature rose through 3 K. What was the specific heat capacity of the mass?
7 How much heat would be needed to melt 0.5 kg of ice at 0°C?
8 How much water at 100°C would boil away when it absorbed 1 MJ, (10^6 J), of heat energy?

What have you learnt?

1 Do you know the difference between the specific heat capacity of a substance, and the thermal capacity of an object?
2 Do you know how temperature can be measured?
3 Can you explain the influence of seas on the climate of nearby land masses?
4 Can you give some uses of water resulting from its large specific heat capacity?
5 Do you know why energy is needed to melt a solid, or evaporate a liquid?
6 Do you know how molecules store heat energy in a solid object?
7 Do you know what is meant by 'dynamic equilibrium'?
8 Do you understand the terms 'latent heat of fusion' and 'latent heat of vaporization'?

⬡12 **Waves**

Objectives ────────────────────────

After reading this chapter you should
- know that waves transfer energy
- know the difference between longitudinal and transverse waves
- know what determines the frequency and speed of waves
- be able to apply the wave equation
- understand wave behaviour in reflection, refraction, diffraction and interference.

12.1 Introduction

When most people think of waves, they think of water waves. People in coastal areas know the damage that large water waves can do to coastlines, breakwaters, or sea walls (Fig. 12.1). It is a property of waves that they **transfer energy**.

Waves have other qualities indicated by Figs. 12.2 and 12.3. Any regular wave has a **wavelength** associated with it, being the distance, λ, between neighbouring wave crests (the highest parts) or the distance between neighbouring troughs (the lowest parts).

The waves travel at a certain **speed**. Ordinary water waves travel at greater speeds in deeper water, but other kinds of wave can travel at very different speeds. Usually the speed of waves in a given material depends on the material.

12.2 The slinky

A convenient way of seeing some other wave properties is to use a long spiral of metal called a slinky, which is a kind of stretched spring. Imagine this stretched out across a smooth polished floor, fixed at one end and held by hand at the other end.

If you move your hand rapidly from side to side while holding the slinky, you can see pulses of disturbance running along the spring. While the hand moves from side to side, the pulses run in a perpendicular direction along the slinky. When the **displacements in a wave are at right angles to the direction of wave motion**, the waves

Figure 12.1 Large waves breaking against the sea wall at Lyme Regis, Dorset (Richard Austin)

are called **transverse waves**. Water waves are called transverse, because the surface of the water is displaced vertically while the waves travel across the surface horizontally.

Another kind of wave may be sent along the slinky. When you move your hand backwards and forwards along the line of the spring, you see a series of compressions and decompressions of the spring moving uniformly along it. This time none of the spring is displaced sideways. Instead, each coil of the spring is displaced alternately backwards and forwards along the line of the spring. When the **displacements in a**

Figure 12.2 Gentle water waves of wavelength λ approaching a beach

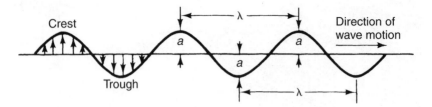

Figure 12.3 *the amplitude a and wavelength λ of transverse water waves are shown. The vertical arrows show by how much the water's surface is displaced*

wave are parallel to the direction of wave motion, the waves are called **longitudinal waves**. Compression waves are longitudinal.

In both transverse and longitudinal waves the disturbances pass along the spring transferring energy as they go; but the **substance as a whole does not move along when waves pass through it**. It just oscillates to and fro. This is a general property of waves.

In Fig. 12.3 the **maximum displacement at a point** in the path of a wave, **measured from its undisturbed position**, is called the **amplitude, *a*,** of the wave. Waves with more amplitude transfer more energy than smaller waves, a point well understood by those who would like to convert the energy in sea waves into electrical energy for use on shore. The attraction of sea waves is partly that this form of energy is renewable, but also that the biggest sea waves tend to occur in winter when there is the greatest demand for electricity.

12.3 The wave equation

If you look at Fig. 12.3 you can see that, when any part of the spring oscillates once, the wave on the spring moves forwards one wavelength, λ. The **frequency, *f*,** of the waves is the number of waves emitted in one second, so *f* waves pass each point on the spring every second. Frequency is measured in cycles per second, or **hertz**, (Hz). During one second the waves move forward $f \times \lambda$ metres, and so the speed of the waves is *v* metres per second where:

$$v = f \times \lambda \qquad\qquad (12.1)$$

This is called the wave equation, and it applies to all types of wave.

> **Example**
> If a boat at anchor rises and falls with the waves ten times each minute, and the waves are seen to be 18 metres apart, how fast do the waves travel?
>
> **Answer A**
> The wavelength is 18 m, and 10 waves pass by in one minute. So the waves move forward $10 \times 18 = 180$ m in 60 s. Their speed is $v = $ distance gone/time.
>
> $$v = 180/60 = 3 \text{ m/s}.$$

12.4 Seeing waves in a ripple tank

A ripple tank is a tray of shallow water with a clear glass bottom. The tray is supported on legs over a horizontal white screen as in Fig. 12.4. A small lamp is supported above the tray so that, in a darkened room, the shadows of ripples produced in the tray are projected onto the screen below it.

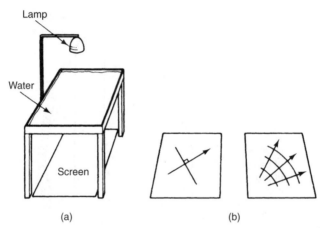

(a)   (b)

Figure 12.4   (a) A ripple tank and (b) shadows of waves appear on the screen below the tank. The arrows are added to show their direction of travel

The edges of the tray are often lined with gauze so that ripples are absorbed and not reflected back across the tank.

Circular ripples can be produced by dipping your finger tip into the water, and straight ripples can be produced with the edge of a ruler. If a barrier is placed in the water, it is possible to see how the waves are reflected from it.

12.5 More wave properties

All waves share some common properties, which are described below.

(a) *Reflection*

Fig. 12.5a shows plane (straight) waves rebounding from a straight barrier, as, for example, in a ripple tank. The direction of travel of the waves is always perpendicular

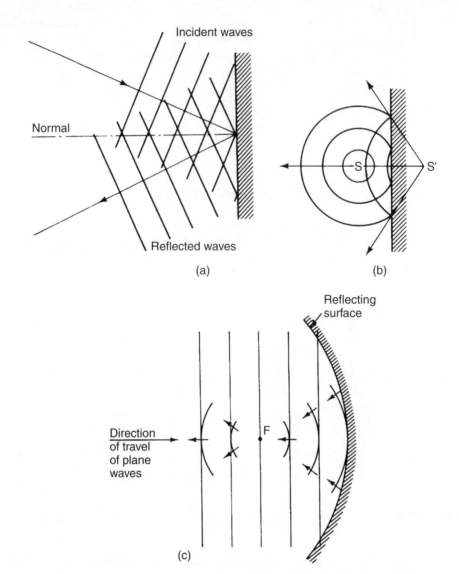

Figure 12.5 (a) The direction taken by incident and reflected waves make equal angles with the normal to the reflecting surface. (b) After circular waves from S are reflected at a straight barrier, they travel as if they had come from S'

to the line of the crests or troughs. The direction of motion of the reflected waves is marked in the figure. Although the reflected waves head in a new direction, neither their speed nor their wavelength is changed by reflection.

Fig. 12.5b shows circular waves rebounding from a straight barrier. Again, the speed and wavelength of the waves are not changed by reflection, but their curvature is changed. The reflected waves are curved as if they had come from S', which is as far behind the barrier as S is in front of it. S' is called the image point of the original source S.

Fig. 12.5c shows plane waves being brought to a focus after reflection at a curved surface. This behaviour is used in the design of radio telescopes and astronomical mirror telescopes.

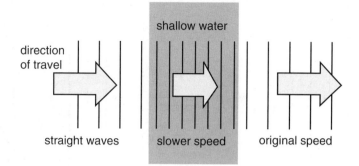

Figure 12.6(a)

Figure 12.6(a) *Straight waves in a ripple tank slow down on crossing a region of shallower water. Their wavelength is reduced when they travel slower*

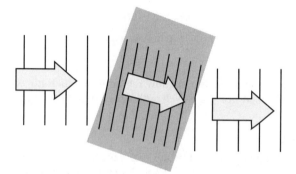

Figure 12.6(b) *On meeting the boundary obliquely, the waves are refracted, and advance in a new direction*

(b) *Refraction*

When waves travel from one medium into another, both their wavelength and speed can be altered, while the frequency of the waves stays the same as when they were emitted. In a ripple tank, refraction can be demonstrated by changing the depth of water in part of the tank by laying a piece of plane glass on the bottom. Fig. 12.6 shows what happens when straight waves are slowed down on passing over shallower water, and then speeded up again over deeper water. The wavelength first decreases, and then increases again.

Fig. 12.6b shows the waves meeting boundaries obliquely, and travelling in new directions when they change their speeds. Chapter 15 gives more details about refraction.

(c) *Diffraction*

A barrier with a small gap in it allows waves in a ripple tank to pass through the gap. Fig. 12.7 shows how waves pass straight on through the gap, but there is evidence that some wave energy at each end of the wavefronts deviates off into the shadow region behind the barrier. This effect, called **diffraction,** is most obvious when the width of the gap is close to the wavelength of the waves. When the width of the gap is much larger than the wavelength, there is little obvious deviation of the wavefronts.

Diffraction also happens when waves pass round small obstacles approximately one wavelength across. But when the obstacle is much larger than the wavelength, the

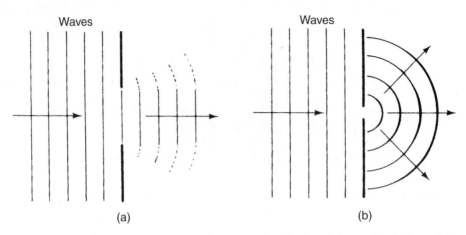

Figure 12.7 (a) Waves passing through a gap much wider than their wavelength show slight diffraction effects. (b) Waves passing through a gap comparable in width with their wavelength show marked diffraction effects

diffraction is much less and there is a shadow region into which no waves are transmitted. Neither the wave speed nor the wavelength are changed by diffraction.

(d) *Interference*

When waves of one frequency pass through two gaps in a barrier, they create two sets of circular waves in a ripple tank. Fig. 12.8 shows the pattern of waves where they cross one another. At places where two crests are superimposed there are larger crests; and where two troughs combine, there are larger troughs. As the waves move forward, the big troughs replace the big crests and vice versa, so paths are created where most of the energy is channelled. This is called **constructive interference.**

Where the waves interfere constructively, they arrive at a point in step with one another which results in waves of large amplitude. The waves are said to be **in phase** at this point. They may have travelled identical distances from identical sources to be in phase, but also their path lengths may **differ by a whole number of wavelengths** for this condition to occur. This explains why there are several paths where constructive interference occurs.

Between these paths are places where the crests from one source of waves are superimposed on the troughs from the other source. This gives rise to waves of minimum amplitude and little energy, where the superimposed waves combine to cause **destructive interference.** The waves at these points are said to be **out of phase.** Travelling from identical sources, the waves have path lengths which **differ by an odd number of half wavelengths** for destructive interference to occur (see Fig. 12.8).

Questions

1 Television aerials are rods supported at right angles to the direction of the transmitter. Do television aerials respond to transverse or longitudinal waves?
2 What property of waves is being used when satellite transmissions are focused by a dish aerial?
3 How should a satellite dish aerial be positioned in order to receive the strongest signals?

⬡13 Sound waves

13.1 Making sound waves

Sources of sound cause pressure variations in the air or some other substance. A loudspeaker, a stretched string, or a drum can cause sounds when they vibrate, because their vibrations alternately compress and decompress the layers of air molecules nearby. These disturbances travel away from their source and cause vibrations in our ears and the sensation of sounds.

Mechanical systems like a stretched string have a natural frequency of vibration which typically depends on their mass and stiffness. A thick guitar string vibrates less frequently and produces notes of lower pitch than a thinner string at the same tension.

3.2 Sound needs a substance

an electric bell is sounded inside a bell jar, the sound is not heard once the air inside jar has been pumped out. You cannot create pressure variations where there is no stance to compress. So **sounds cannot be produced in a vacuum, and sound**

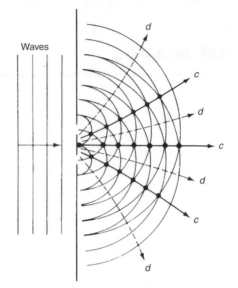

Figure 12.8 An interference pattern of waves emerging in phase from two identical silt sources. Paths marked 'c' indicate constructive interference and those marked 'd' indicate destructive interference

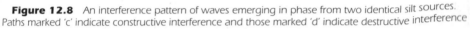

4   Radio waves of wavelength 1500 m have a frequency of 200 kHz. How fast do the waves travel?
5   Long wavelength radio waves may be received in valleys remote from a direct line to the transmitter. What property of waves enables this to occur?
6   (a)   What is happening when waves from two sources cancel each other out?
    (b)   Where does their energy go?
7   Water waves pass into an area where the water is deeper. What will happen to:
    (a)   their speed, and (b) their wavelength?
8   What sort of sea water waves are likely to transfer most energy?

What have you learnt?

1   Do you know the difference between transverse and longitudinal waves?
2   Can you explain the meanings of wavelength, frequency and amplitude applied to waves?
3   Can you write down the wave equation and explain what it says?
4   Do you understand how straight and curved waves are reflected at a str barrier?
5   Can you explain what happens to waves when they are refracted?
6   Do you know how waves can be diffracted?
7   Do you know the difference between constructive and destructive ence of waves?
8   Do you know how the energy in a wave is related to its amplitude?
9   Do you know what determines the speed of waves?
10 Do you know that the frequency of waves is determined by their s

will not travel through a vacuum. Astronauts on the Moon's surface could not speak to one another without using radio waves, as there is no atmosphere around the Moon to carry sound waves from the speaker to the listener. A two-way radio in each space suit must convert sound waves into radio signals, and be able to receive radio signals and convert them into sound waves.

13.3 Describing sound waves

Fig. 13.1 indicates the way a loudspeaker cone vibrates, and shows the flow of sound waves away from their source. The cone moves alternately forwards and backwards along the direction of travel of the sound waves. So the layers of air are displaced alternately forwards and backwards along this direction. This means that **sound waves are longitudinal compression waves** (see chapter 12).

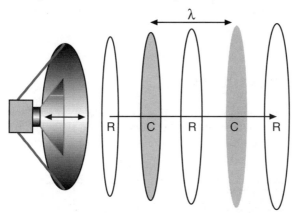

Figure 13.1 *A loudspeaker cone moves left and right alternately. This causes compressions, C, and rarefactions, R, which move from left to right away from their source. The distance between the compressions is the wavelength of the sound in air*

When the loudspeaker cone moves forwards, the layers of air in front of it are compressed and a rise in air pressure, or **compression,** is transmitted through the air. When the cone moves back to complete its vibration, the layers of air in front of it are decompressed and a fall in air pressure, or **rarefaction,** is transmitted through the air. So one vibration of the source generates one cycle of compression and rarefaction in the air.

13.4 Frequency and pitch of sound

The frequency of sound is the number of vibrations reaching a point in one second. Sound reaching our ears vibrates membranes, or eardrums, and causes electrical impulses which nerves carry to the brain (see chapter 14). We hear sound when our eardrums vibrate between about 20 and 20 000 times per second, which tells us the **range of audible frequencies, i.e. from about 20 Hz to 20 000 Hz.** Older people tend not to hear the highest pitched sounds in this range. Sound vibrations of higher frequencies are called **ultrasonic** vibrations, and although human ears cannot detect such high frequency sounds, some animals can hear ultrasonic sounds.

There is a direct connection between the frequency of sound vibrations and the pitch of the note we hear. The deep notes of a bass voice have a low pitch and a low frequency, whereas the top notes of a soprano singer have much higher pitch and frequency. **The pitch of a musical note is determined by its frequency.**

13.5 Loudness

The sound waves of louder sounds carry more energy than for softer sounds. This is because the pressure variations in louder sound waves have larger **amplitude**. The amplitude is the difference between the maximum (or minimum) pressure in the wave and the undisturbed atmospheric pressure. So in general **large amplitude pressure waves transfer more energy than smaller amplitude waves**.

The loudness of sound, measured in **decibels,** (see section 14.5), can be registered with a hand-held decibel meter, containing an in-built microphone.

13.6 Using microphones

A microphone is a sound detector. It responds to the energy in sound waves by converting it into electrical impulses of the same frequency, and these can be displayed on a **cathode ray oscilloscope** (see Fig. 13.2a). The trace displayed on the CRO can be

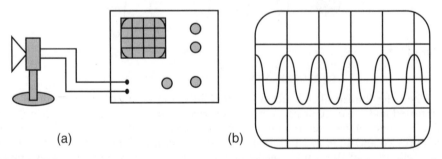

      (a)   (b)

Figure 13.2 (a) A microphone converts sound waves into electrical signals which are displayed on the cathode ray oscilloscope. A typical trace is shown in (b). Five cycles are displayed across the screen

used to measure the frequency of a note, since it shows how the electrical signal from the microphone varies with time. The distance across the tube horizontally, which contains one cycle of the waveform, is a measure of the time for one cycle. The frequency, f, is calculated using the formula:

$f = 1$/time in seconds for one cycle   (13.1)

So, for example, if the timebase of the CRO in Fig. 13.2b is set at 5 ms/division, then we can see that one cycle takes five milliseconds, and so the frequency is $f = 1/0.005$ s $= 200$ Hz.

We can also use this apparatus for comparing the amplitudes of sound waves. A louder sound will produce a trace with larger amplitude in the vertical y direction. We can also study the shape of the sound waveform which, as we shall see in chapter 14, has different characteristics for different musical instruments, and for different voices.

Microphones can detect ultrasonic as well as audible sounds and a CRO can display them. Alternatively a microphone can be connected directly to a frequency meter which counts the cycles of the electrical signal it receives in unit time, and displays the frequency of the sound in cycles per second, or hertz.

Self test

(a) If each sound wave takes 1/500 s to pass a detector, what is the sound's frequency?
(b) How long does it take one wave of frequency 100 Hz to pass the detector?

((a) 500 Hz, (b) 1/100 s)

13.7 Sound waves cause no flow of matter

It is a general result that when waves carry energy through a material, there is **no flow of matter along with the waves**. Sound waves do not create a flow of air away from their source, since each layer of air is displaced both forwards and backwards about a mean position. **Energy is transferred, but mass is not**.

In the path of a compression, first one layer of air is displaced, and then another layer which is further from the source. Their displacements are not simultaneous, nor are they maintained. A moment later the first layer is displaced back towards its original position and then the other layer.

13.8 Speed of sound in air

Two microphones are set up so that sound reaching the first will start a millisecond timer and sound reaching the second microphone will stop the timer. The timer is placed midway between the two microphones as in Fig. 13.3.

A loud clap is made near the first microphone. This starts the timer. When the sound reaches the second microphone, the timer stops and shows that the sound took perhaps 20 ms to travel the distance of 6.4 m between the microphones. We can find the speed of the sound wave.

Speed of sound in air = distance gone/time taken = 6.4 m/0.020 s = 320 m/s   (13.2)

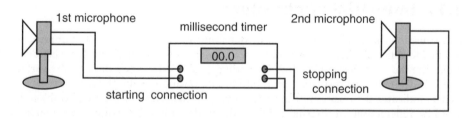

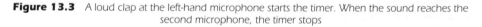

Figure 13.3 A loud clap at the left-hand microphone starts the timer. When the sound reaches the second microphone, the timer stops

13.9 Accuracy of measurements

Each of our readings, (that is 6.4 m and 20 ms), were found to **two significant figures**. Notice that 6.4 does not mean 6.40, which is given to three significant figures. Experimental results can only be given to a certain level of precision or accuracy, and the calculated result cannot be regarded as any more accurate than the initial readings. Our results suggest that the speed of sound in air was likely to be more than 310 m/s and less than 330 m/s.

Beware of quoting too many significant figures when you use a calculator. If the distance had been 6.1 m and the time 19 ms, you would calculate the speed of sound as 321 m/s, ignoring all the decimal places indicated by your calculator. Having shown this figure, you then round it down and write '= 320 m/s (2 sig. figs.)'. This shows that you understand that your readings were only reliable to two significant figures.

The speed of sound in air can also be established using an echo technique, in which a timer is started when sound is emitted, and the returning echo stops the timer after the sound waves have travelled to a wall and been reflected back towards their source. The distance gone in the timed interval is the distance from the timer to the wall plus the distance back from the wall to the timer.

13.10 Echo sounding

The echo method is used at sea to find the depth of water beneath the hull of a ship, or perhaps the depth of a shoal of fish beneath the ship. Pulses of sound are emitted rather than a continuous train of waves, so that the time for the echo can be measured. The speed of sound in sea water is greater than its speed in air, and this must be allowed for in the calculation of the required depth:

Depth below the hull = speed of sound in sea water $\times$ timed interval/2   (13.3)

Self test

A car backfires, and its echo reaches the driver 120 ms later.

(a) If the speed of sound in air was 320 m/s, how far had the sound travelled?
(b) How far was the car from the sound reflector when it backfired?

((a) 38 m, (b) 19 m [2 sig. figs.])

13.11 Industrial applications

Another use of the echo technique is in detecting internal cracks, discontinuities or faults inside solids having uniform cross-section. A pulse of ultrasound travels down through the material (a solid steel girder perhaps), and is reflected back to the surface where it is detected and displayed on a CRO. The position of the pulse on the face of the CRO indicates the position of the reflecting bottom surface of the steel. If the pulse were to be reflected at an internal crack in the girder, the echo would return sooner, and another peak would be displayed to the left of the first one. In this way faults inside the steel may be located (Fig. 13.4).

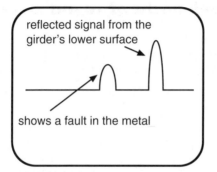

reflected signal from the girder's lower surface

shows a fault in the metal

Figure 13.4 *A CRO trace shows the larger reflected pulse of ultrasound from the bottom of a girder, and the reflection from an internal fault in the metal*

13.12 What determines the speed of sound?

The speed of sound varies according to the material in which the waves travel, so there is one speed for steel, and another for copper, etc. In air the speed varies with the temperature, being faster when the temperature of the air rises. Kinetic theory tells us that **gas molecules move faster on average at higher temperatures,** so it is not surprising that hot air can transmit compression waves faster.

However the **speed of sound is less in gases with larger relative molecular masses,** (RMM), than in gases like hydrogen with low RMM values. Kinetic theory explains that at a given temperature the kinetic energies of all gas molecules have the same average values. This means **the more massive molecules move more slowly on average than the less massive molecules.** So, understandably, the speed of sound in carbon dioxide is less than in nitrogen, and the speed in nitrogen is less than in hydrogen at the same temperature. (The RMM values are 44 for CO_2, 28 for N_2, and 2 for H_2). In explaining these results we have used the formula obtained in Appendix A for the kinetic energy of a molecule:

$$KE = (1/2) \times mass \times (velocity)^2, \text{ or } (1/2) \, m.v^2 \tag{13.4}$$

—— **Self test** ——

Explain why sound waves travel more slowly at an altitude of 10 000 m above sea level than their speed at 10m above sea level.

(Think about the temperature).

13.13 Thunder and lightning

The difference in speeds of light and sound in air creates the time delay between seeing a lightning flash in a thunder storm and hearing the corresponding clap of thunder. Light travels so fast in air that we experience the flash only a minute fraction of a second after the electrical discharge occurs some kilometres away. A few seconds later we may hear the thunder. From a measurement of the delay, we can estimate how far away the electrical disturbance took place. If the delay was 5 s, and if sound can travel in air at 320 m/s, then the sound travelled 5 × 320 metres before it reached us, i.e. it was emitted 1600 m away.

13.14 The wavelengths of sound waves

From measurements of the speed of sound waves and the frequency of their source, we can find their wavelength. For example, if the speed of sound in air is 320 m/s, and a note of frequency 240 Hz is played, we can apply the wave equation and say:-

$v = f \times \lambda$, where λ is the wavelength (see Equation 12.1), so
320 m/s = 240 Hz $\times \lambda$, and
λ = 320/240 = 1.3 m approximately.

While waves with large frequency have a small wavelength, waves with low frequency have large wavelengths in air up to about λ = 8 m where f = 40 Hz. **The wavelength is the distance between two neighbouring compressions**, or between two neighbouring rarefactions (decompressions).

Self test

Show that if the frequency of sound in air is f = 3.2 kHz, its wavelength is about 10 cm.

13.15 Sound and the general properties of waves

We have already referred to echoes which occur when **sound waves are reflected** at the wall of a building. It is not easy to show that the waves follow tracks like those of snooker balls when hitting a cushion. Sound waves are easily scattered. Nevertheless, where measurements can be made, sound waves do have this property (see Fig. 13.5).

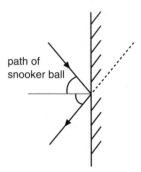

path of
snooker ball

Figure 13.5 A snooker ball and sound waves follow similar paths

Refraction of waves occurs when the waves travel into a new substance, in which they move at a new speed. This is what geologists find when they study the layers of rock in the Earth's crust by detonating an explosive charge in the earth, and detecting the signals reflected from different layers below them. The reflected signals give information about the underlying rock layers, and the technique is used by mining engineers and oil prospectors.

The **refraction of sound** occurs, for example, when a gas-filled balloon is held in front of a small loudspeaker emitting sounds in air. If the balloon is filled with carbon dioxide, it refracts the sound rather like a converging lens refracts light emitted from a nearby source (see chapter 15).

The trunk of a beech tree may be more than one metre in diameter, and the wavelength of sound when you shout may be very similar. This is the condition in which the sound waves most readily **diffract** around the obstacle, and so your voice may be heard on the far side of the tree. At night however, light from your torch or lantern would cast a dark shadow behind the tree. The wavelength of light, as we shall see, is very small compared with the diameter of the tree trunk. So this is not a favourable set up for demonstrating the diffraction of light.

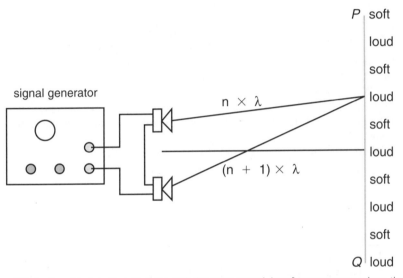

Figure 13.6 Two loudspeakers in series emit the same sound. Interference occurs where the sound waves overlap, and loud and soft sounds are detected along a line PQ

Fig. 13.6 shows a signal generator driving two identical loudspeakers. Sound waves of a certain frequency spread out into the space to the right of the speakers where they overlap. This produces the **interference effect** discussed in chapter 12 for water waves. The waves are superimposed so that they reinforce each other at some places and cancel each other out at other places. The sound is alternately loud and soft when detected along the line PQ in the figure.

The condition for a loud sound would be that the two sets of waves arrive in step, or **in phase** with one another. This would be like two crests of water waves arriving together to make a big crest. In sound it might be two compressions arriving together followed by two rarefactions arriving together. The distances from a loud point to the speakers must differ by a whole number of wavelengths for this to occur. Remember zero is a whole number.

A soft sound occurs where the waves have travelled distances which differ by an odd number of half wavelengths. They arrive at the point **out of phase**, and destructive interference occurs.

1   One end of a ruler is clamped to a table, and the free end is 'twanged'. Why does it make a sound?
2   Explain the change in sound when the same ruler is clamped near its mid-point and 'twanged'.
3   If the ruler vibrates once every 0.01 s, what is its frequency of vibration?
4   How is the wavelength, λ, of sound waves related to their frequency, f, and speed, v, in air?
5   Write down four general wave properties which can be demonstrated with sound waves.
6   In a thunder storm, thunder is heard 4 s after a lightning flash.
    (a) Why is there a time delay between seeing the lightning and hearing thunder?
    (b) If sound travels at 325 m/s in air, how far away was the lightning from the observer?
7   Fig. 13.7 shows four displays on a CRO connected as in Fig.13.2a to a microphone. The CRO has a timebase setting of 5 ms/division.
    (a) Find the time for one cycle of the sound in Fig. 13.7a.
    (b) Which display indicates sound with the lowest pitch?
    (c) Which display indicates the loudest sound?
8   Explain why some dog whistles are heard by dogs, but not by people nearby.
9   Why does sound travel slower in carbon dioxide than in air at the same temperature?
10 Explain why sound waves do not reach us from the Sun.

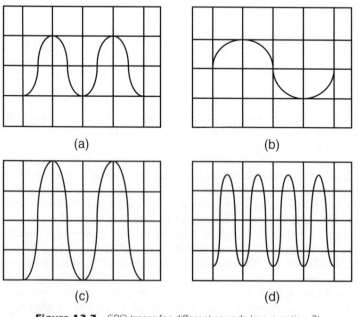

(a)   (b)

(c)   (d)

Figure 13.7 CRO traces for different sounds (see question 7)

1   Do you know that sound waves contain alternate compressions and decompressions?
2   Do you know that sound is a longitudinal wave motion?
3   Do you know why sound will not pass through an evacuated space?
4   Can you relate the time a sound wave takes to pass you, to the frequency of its sound?
5   Do you know that sound travels at different speeds in different materials?
6   Can you describe a method for finding the speed of sound in air?
7   Can you relate the speed of sound waves to their wavelength and frequency?
8   Do you know what frequencies of sound may be heard by humans?
9   Do you know that ultrasonic waves have higher frequencies than we can hear?
10   Do you know that some microphones respond to ultrasonic waves?
11   Do you know that loudness, which is related to the amplitude of sound vibrations and the energy they transfer, is measured in decibels?
12   Can you explain why two loudspeakers emitting the same musical note may sound louder than one loudspeaker in some places, but not as loud as one loudspeaker in other places?

14 The sounds we hear

— Objectives —

After completing this chapter you should
- understand how the ear hears sounds
- know what frequencies of sound are normally audible
- know what forced oscillations are
- understand resonance
- be able to describe how musical instruments produce their sounds
- be able to describe the fundamental and overtones in a musical sound
- understand why different instruments have different tone or quality
- know what is meant by noise.

14.1 The human ear

You might imagine that the most important part of the ear is the part you can see. But this part only receives sound waves and directs them into the ear hole. The hearing part of the ear is at the inner end of a channel inside the skull.

The sound receptor is a membrane called the **ear drum,** which vibrates as the compressions and rarefactions in sound waves arrive (Fig. 14.1). Three linked bones called the **hammer, anvil**, and **stirrup** act as levers and transmit the movement of the ear drum to another membrane called the **oval window**. Beyond this is a fluid contained in the **cochlea** in the inner ear.

The coiled cochlea contains a series of **receptor cells** along its length. Vibrations in the fluid of the cochlea stimulate the receptor cells which transmit signals to the auditory (hearing) centre of the brain, and you hear the sound.

14.2 Hearing notes of different frequency

High pitched notes have many vibrations per second. Beyond the top end of the piano keyboard there is an upper limit to the frequencies of sounds we can hear (see chapter 13). Young people may hear notes of frequency as high as 20 000 Hz, or 20 kHz, but in later life people find they can only hear sounds with frequencies lower than say, 12 kHz, or some smaller rate of vibration.

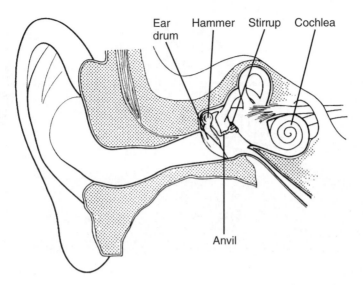

Figure 14.1 *A simplified diagram showing the main parts of a human ear*

Prolonged exposure to loud sounds can damage your hearing permanently. The ear is a delicate mechanism, and hearing can be restricted for a variety of reasons. Operators of very noisy machines in industry are required to wear protective ear muffs which restrict the intensity of sound waves entering their ears. Music whether heard at pop concerts or from Walkman earphones may also be loud enough to cause gradual and serious loss of hearing over a period of time.

14.3 Resonance

An object which can vibrate has a natural frequency at which it will do so. For example, a child's swing or a pendulum when given an initial push, will swing to and fro with a definite frequency. All mechanical systems have their own natural frequencies of vibration.

An object may be forced to vibrate at a variety of frequencies, as for example the wooden soundbox of a violin, viola or cello. Suppose for example the metal bodywork of a car were forced to vibrate at the frequency of its internal combustion engine. It would be possible for the frequency of the engine's forced oscillations to match the natural frequency of the car's body. This would result in larger amplitude vibrations in the body of the car. It is an example of resonance; we say the car's body resonates with the engine. **For resonance to occur, the frequency of forced vibrations must match the natural frequency of the system being forced**.

Resonance can be disastrous. An example was the destruction of the Tacoma Narrows Bridge in the state of Washington, USA in 1940 (see Fig. 14.2). A strong transverse wind produced impulses on the suspension bridge which built up large resonant oscillations. After some hours of oscillation, the bridge collapsed. Nowadays extensive tests on models of structures in wind tunnels ensure that the dangers of resonant vibrations are avoided when the fullscale structure is built.

Similar tests are carried out on new aircraft. If the natural frequencies of each part of the aircraft are established, steps can be taken to ensure that the engine vibrations do not cause dangerous resonant conditions which could harm the structure when in flight.

Figure 14.2 The Tacoma Narrows bridge, Washington State. The roadway is oscillating in a cross-wind shortly before its collapse

14.4 Musical instruments

We shall conveniently divide the instruments into three categories:

(a) stringed instruments
(b) percussion instruments
(c) wind instruments.

(a) *Stringed instruments*

A string is made to vibrate either by plucking (as in the guitar or harp), striking (as in a piano) or bowing (as in a violin, viola, cello or double bass). In each example the string vibrates at a certain frequency.

In the harp, the string is the sole source of the sound. It causes the surrounding air molecules to vibrate to and fro at the same frequency, sending sound waves into the air.

In the piano a sound board is mounted in a plane parallel to the strings. It vibrates in sympathy with the strings, amplifies their sound and transmits it to the air nearby. When the piano is played, the strings vibrate after being hit by felt-covered hammers.

Fig. 14.3 shows the construction of a violin. A string is either plucked or bowed at a point between the bridge and the fingerboard. The pitch of the note depends on the tension in the string which is adjusted during a tuning up period. When a string is bowed, the pitch is varied by changing the string's length from the bridge to the point where it is held against the fingerboard. Higher frequencies and pitch are obtained when the string is made shorter or has greater tension.

Vibrations of the strings are transmitted through the bridge to a sound post inside the hollow body of the instrument. This causes the **body to vibrate in sympathy with the strings**, and the sound is amplified and transmitted into the surrounding air. The body of a guitar has a similar amplifying effect.

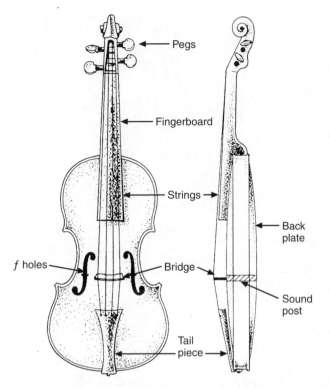

Figure 14.3 The main parts of a violin

Labels: Pegs, Fingerboard, Strings, Back plate, f holes, Bridge, Sound post, Tail piece

(b) *Percussion instruments*

This family of instruments includes all types of **drum, the cymbals and the triangle**. They are struck to produce sound. Many other percussion instruments can be found in a full orchestra, but only these three are considered here.

The stretched membrane of a drum vibrates according to its tension, but the sound also depends on where the membrane is struck. The frequency of vibration of the membrane determines the pitch of the sound, which is **amplified by the vibrations of the air enclosed inside the body of the drum**.

The cymbal is a circular metal plate, supported at its centre but free at its edges. It may be hit, for example, with a drum stick near its edge, or two cymbals may be struck against each other. The vibrations of the metal plate are complicated, and they produce a sound of indefinite pitch.

The triangle is a metal rod bent to form a triangular shape. It is struck by another metal rod and produces a variety of high pitched notes.

(c) *Wind instruments*

If you blow across the end of a hollow pipe, you may hear a sound of definite pitch. **The air column inside the pipe has a natural frequency** at which it will vibrate when you blow. The **flute** and **piccolo** illustrate this effect as in Fig. 14.4a.

In the **clarinet** and **saxophone** (Fig. 14.4b) **a single reed vibrates**, supported by a solid mouthpiece. The reed's vibrations are coupled to a **resonating column of air** in a tube. When the tube is effectively made shorter by opening holes along its

Figure 14.4 (a) flute, (b) clarinet, (c) oboe, (d) bassoon, (e) trumpet, (f) trombone

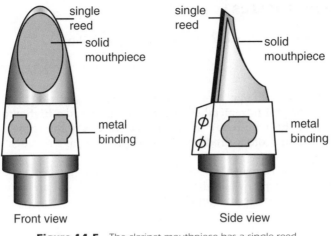

Figure 14.5 The clarinet mouthpiece has a single reed

sides, the pitch and frequency of the sound are raised. See Fig. 14.5 for the single reed mouthpiece.

In the **oboe** and **bassoon** (Fig. 14.4c and d), the mouthpieces consist of **two pieces of reed** through which the player blows air into a coupled air column. As **the air column resonates** with the vibrations of the reeds, sound waves are emitted from the open end of the tube. The pitch of the note may be raised by opening holes along the sides of these instruments, just as with the single reed instruments. Fig. 14.6 illustrates double reeds.

A number of **brass instruments have no reeds: a trombone, tuba, trumpet,** and **French horn,** for example, have none. In these instruments the vibrations are produced by the players' lips, held against a small cup-shaped mouthpiece (see Fig. 14.4e). The vibrations are again coupled to resonating air columns with lengths controlled by means of valves, or in trombones by sliding a tube in or out (see Fig. 14.4f). Where no control of the length of the air column is included, the instrument may sound a more limited range of notes, as in a **bugle** call for example.

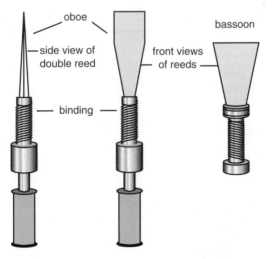

Figure 14.6 The oboe and bassoon have double reeds

14.5 Musical notes

A musical note has a characteristic pitch, loudness and quality.

(a) *Pitch*

We have already related the pitch of a note to its frequency of vibrations. As a rule of thumb it may be helpful to remember that a note **one octave higher on a musical scale has double the frequency of the lower note**. Chords consisting of several notes sounded together often produce effects dependent on the ratios of their individual frequencies.

Some musical instruments are good at sounding low pitched notes, the double bass and tuba for example, while the violin, oboe and piccolo are good at sounding notes of higher pitch (but see later under 'Quality of sound').

(b) *Loudness*

The loudness of a sound depends on the ability of the ear to 'hear', which depends on the movement of the ear drum as sound waves arrive. The bigger the movement of the ear drum, the stronger will be the signals sent to the brain, and the louder the sound we hear.

Loudness can be measured by the rate of energy flow as sound waves pass a detector or microphone. On one scale of measurement the faintest sound we can hear is estimated as an energy flow rate of 10^{-6} μW/m^2. This threshold of hearing is rated as a sound level of 0 decibels (db). Table 14.1 lists some sound levels in decibels and their corresponding intensities in microwatts per square metre.

(c) *Quality*

Most people have no difficulty in telling that a violin and oboe sound differently even when they sound the same note. The difference is not in the pitch of this fundamental note, but in the other notes that each instrument sounds at the same time. These other

Table 14.1 Intensities and sound levels of some sounds

Sound		Intensity (μW/m^2)	Sound level (db)
Silence	Threshold	10^{-6}	0
	Faint whisper	10^{-4}	20
Quiet	Turning page of newspaper	10^{-3}	30
	inside of train (windows shut)	10^{-1}	50
Moderate	Conversation	1	60
	Heavy traffic	10^2	80
Loud	Loud motor horn, or pneumatic drill	10^3	90
	Thunder, jet plane taking off	10^5	110
Very loud	Unacceptable pain	10^6	120

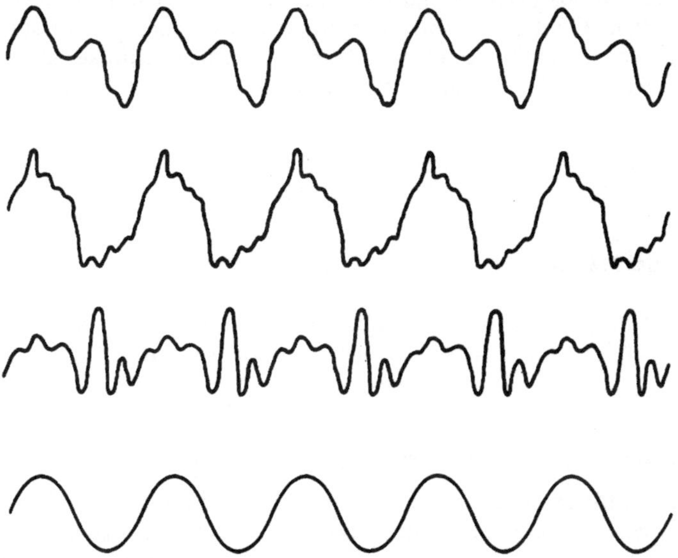

Figure 14.7 *The notes have the same frequency but different quality*

notes which have higher pitch than the fundamental are called **harmonics**. Different instruments produce different harmonics with intensities in different ratios to their fundamental. This means that the sounds played on different instruments are different, and enable the instruments to be distinguished.

If musical sounds are picked up by a microphone and the resulting signal displayed on an oscilloscope (Fig. 14.7), the differences are clear. A tuning fork emits an almost pure note of one frequency with little energy in harmonics. The waveforms of the same note played on other instruments show the combined effects of the harmonics and their fundamental notes. While none of the sounds can be regarded as better than another, they are said to have **different qualities**.

People who cannot hear high pitched notes very well may not properly hear all the harmonics when an instrument is played. This will give the sound a different quality from that heard by a person who can hear the higher notes. The top note on the piano keyboard has a fundamental frequency of about 4 kHz.

14.6 Recorded music

Good audio equipment must accurately reproduce all the fundamentals and all the harmonics being sounded when instrumental music is recorded. This includes frequencies up to around 20 kHz, even though some listeners may be unable to hear the highest of these frequencies.

14.7 Noise

Often people use the word 'sound' for something they want to hear, and 'noise' for what they do not want to hear.

In general, musical sounds are made up of a certain limited number of frequencies.

They are regarded as sounds even though some people may not want to hear them. Motor traffic, aircraft in flight and trains all produce a complex range of sounds of many unrelated frequencies at the same time. This is described as **noise**. To a physicist or sound engineer, noise is a random mixture of sounds of different frequencies and amplitudes.

14.8 Noise control

Many people believe that there is too much noise about – produced by machines or by other people. What can be done to reduce the level of noise we hear?

(a) *Reduce the amount of noise made by a source*

In a motor car engine, for example, it is a priority to balance the crankshaft so that it does not vibrate when it rotates. The next aim is to make sure that metal panels surrounding the crankshaft do not vibrate in resonance with its rotations. Finally it is important that the base on which the engine rests cannot vibrate like a sound board in sympathy with the engine. The engine is usually fixed on anti-vibration mountings which have blocks of rubber or equivalent material between the engine and the mounting base.

A great deal of noise is created in any internal combustion engine when exhaust gases from the cylinders are released into the surroundings. The release of gas at high pressure produces a noise which can be unpleasantly loud. The noise is greatly reduced by using one exhaust pipe to transfer exhaust gases from the cylinders to the surroundings, and by placing a so-called **silencer** in the pipe. The silencer is a box wider than the exhaust pipe, containing various internal plates. Its function is to reduce the variation in gas pressures as the exhaust gases reach the open end of the pipe. The exhaust noise is therefore substantially reduced.

For people who live near airports, the **jet engine is an important source of noise.** Most of the noise is caused by the fast moving exhaust jet flowing past the stationary surrounding air. The high speed of the exhaust gas past the stationary air produces a 'shearing' effect and associated noise. All military jet aircraft and the Concorde passenger aircraft have engines of this type, called **turbo-jets.**

> The **turbo-fan engine is significantly quieter.** Fig. 14.8 shows the rear view of a Rolls-Royce RB 211 engine mounted on the wing of a wide-bodied airliner. It has an air intake fan of diameter about 2.2 m, which is larger than a simple jet engine. The central core of the engine, where the fuel combustion takes place, has a diameter of about 1.0 m. From there the air emerges hot and fast in the usual way. But the exhaust gas is surrounded by a sleeve of more slowly moving air with which it mixes. The result is smoother mixing and less shearing between the jet exhaust and the outside air, and a much quieter

engine. You can see the central core and the edge of the surrounding sleeve in the picture. If there is ever to be a Concorde Mark II, people say it will have to be powered by supersonic turbo-fan engines, to reduce noise (but no such engines exist yet).

Figure 14.8 The rear view of a turbo-fan RB 211 jet engine (Courtesy of Rolls-Royce plc)

(b) *We want to insulate the hearer from whatever unwanted noise is around*

The walls of a building must vibrate very little in sympathy with sound waves that strike them, or ground tremors due to traffic. The conventional cavity wall construction fares no better as a sound insulator than a solid wall of the same thickness. Double glazed windows are good thermal insulators, and they can also be good sound insulators. The airspace between the panes of glass can be at reduced pressure, or alternatively, the panes may be separated by an air gap of at least 100 mm for good sound insulation. The panes should also be firmly mounted.

(c) *The space where the listeners are situated must contain the right materials*

This applies as much to private living rooms as to concert or lecture halls. Suppose a hall is to be used for concerts. The music must be heard clearly. Therefore, the effect of echoes must be controlled. Sounds coming in from outside the hall must be eliminated or become inaudible. Consider the hard surfaces in the hall, like flat stone walls or large panes of glass. They are very good reflectors of sound, and add to the reverberation time during which sound waves die away. Heavy curtains, however, are good at absorbing unwanted sound and are often used for this effect.

People in concert halls absorb sound quite effectively, and the chairs are designed so that they too absorb sound equally well. So the sounds which are heard are not effected by the number of people present in the audience. One hopes the chairs were also designed for comfort!

Questions

1 How does a sound travel through air?
2 What is the range of audible frequencies of sound waves a normal ear can hear?
3 What controls the natural frequency of a vibrating string?
4 (a) What vibrates with 'forced oscillations' in a violin?
   (b) How is the instrument's sound output affected by this?
5 Explain what is meant by 'resonance'.
6 When do coupled oscillations produce resonance?
7 How does the sound of a tuning fork differ from that of a violin playing the same note?
8 How do sound waves differ if one is louder than the other?
9 Explain what is meant by noise.
10 What features of a concert hall enable an audience to hear music clearly?
11 What is the frequency of the note two octaves higher than one of frequency 256 Hz?
12 Explain how notes in (a) a stringed instrument, and (b) a trombone are raised in pitch.
13 What sounds may harm one's hearing?

What have you learnt?

1 Do you know how the human ear detects sound waves?
2 Do you know what can cause hearing defects?
3 Do you understand the terms 'forced oscillations', 'natural frequency' and 'resonance'?
4 Can you explain how sounds are produced in musical instruments?
5 Do you understand the terms 'pitch' and 'loudness'.
6 Do you know why a piano and clarinet produce different sounds?
7 Can you explain the difference between sound and noise?
8 Can you explain the advantages of noise control?
9 Do you know what steps can be taken to control noise levels in buildings?
10 Do you know that loudness can be measured in decibels or in microwatts per square metre?

⬡ 15 Light

Objectives

Objectives

After completing this chapter you should
- understand what sources and sensors of light are
- understand that light exhibits the properties of waves
- be able to explain reflection in a plane mirror
- be able to describe images
- understand the words 'real', 'virtual', 'inverted' and 'laterally inverted'
- understand how a change of speed of light leads to its refraction on crossing a surface
- be able to describe dispersion of white light by a prism
- be able to explain the critical angle and total internal reflection in glass fibres
- understand how converging lenses refract light waves
- be able to construct the image position when light has been reflected at a plane mirror, or passed through a converging lens
- understand magnification of an image, and be able to measure or calculate its value.

15.1 Light sources

Luminous objects emit light. The Sun is the biggest source of light reaching the Earth, but many other sources occur on Earth, from a red hot coal in a fire to a white hot tungsten filament bulb conducting electricity. In the filament bulb electrical energy is transformed into both heat energy and light. **All white hot solids emit all the colours of the rainbow**, and as we shall see later, this combination of colours gives us the sensation of whiteness.

A gas at low pressure can also conduct electricity and emit light waves typical of the gas; sodium vapour street lights, for example, emit mostly yellow light, while a number of other gases are used to emit other bright colours. We can split a beam of light into its various colours as Isaac Newton did when he passed the light through a triangular glass prism (Fig. 15.1).

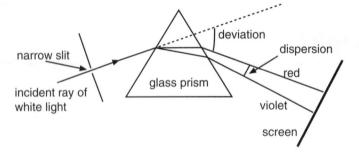

Figure 15.1 A ray of white light is split into colours when refracted by a prism. The red light is deviated less then the violet light. The dispersion is the angle between the emerging red and violet rays

15.2 Light sensors

The **retina** at the back of the eye is a surface made up of light-sensitive rods and cones. These sensors convert the energy they receive into **electrical impulses which are carried by the optic nerve** to the brain for decoding. A normal human eye (see Fig. 15.2a), can distinguish the colours of the rainbow from red to violet and focus on objects ranging from about 30 cm from the eye out to infinity. We refer to very distant objects as 'at infinity'.

In order to control the amount of light entering the eye, there is a coloured iris with a central hole called the pupil of variable diameter through which the light passes. On bright days the pupil is smaller. In a camera, light is admitted through a hole, or aperture, in an adjustable diaphragm.

The light sensor in a camera is a photographic film (see Fig. 15.2b). **Chemical changes occur in the film negative** where it receives light. The image can then be fixed in the film chemically when it is developed, and it can be enlarged and printed onto photographic paper. Photographic negatives are also sensitive to rays we cannot see with our eyes, like X-rays, ultraviolet rays and gamma rays as described in chapter 29. The chemical changes on the negative can also be triggered by bombardment with charged particles. This means that a photographic film can record some radiations we cannot see.

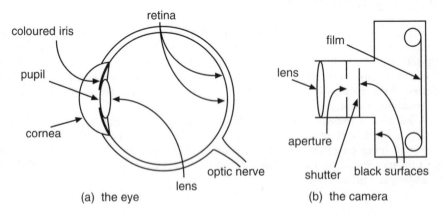

(a) the eye

(b) the camera

Figure 15.2 (a) The eye admits light through the pupil which is surrounded by the iris. Behind the iris is a lens to focus images on the retina. (b) shows a camera in which an aperture admits light through a diaphragm. Images on the film are focused by means of a moveable lens

A video camera senses light images and converts them into electronic signals. These are stored on magnetic video tape. When the image is to be viewed, the signals are converted back into electronic pulses which are passed to a television set where they are displayed again as pictures. The camera can record many images each second, and so it can show both moving and static images.

15.3 Waves and their motion

At the seaside as waves approach the shore, they travel at right angles to their wave-front. Each wave may be parallel to the shore line as it moves into the shallows, and finally breaks into white foam on the shore. The **direction of travel is perpendicular to the wavefronts**.

Light waves also exhibit this property as illustrated in Fig. 15.3. Figures which

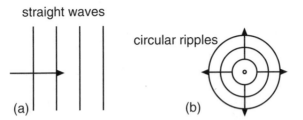

Figure 15.3   Both straight waves and curved waves advance along perpendiculars to their wavefronts

show light rays indicate the directions in which the waves travel. The wavefronts themselves are always perpendicular to the ray directions.

We have seen in chapter 12 that **waves carry energy** and show the properties of **reflection, refraction, diffraction** and **interference.** Since light shows all these properties, we can consider it to be energy transmitted in the form of waves.

15.4 Reflection

In Fig. 15.4, circular waves spread out from a point source at P. After reflection at the plane mirror the waves are still circular, but they are travelling away from P', which is the image of P behind the mirror. As described in section 12.5, P' is as far behind the mirror as P is in front of it, and the line joining P to P' is perpendicular to the mirror line, or **normal** to it.

Fig. 15.5a shows the reflection of a ray at a plane mirror. Rays are narrow pencils of light. The **incident ray** reaches the mirror from its source. The redirected ray which leaves the mirror surface is called the **reflected ray.** A line drawn perpendicular to the mirror at the point where the rays meet it is called **a normal**, and this normal line bisects the angle between the two rays, so angle r in the figure equals angle i.

The law of reflection states that **the angle of incidence equals the angle of reflection:**

angle i = angle r (15.1)

In Fig. 15.5b three rays from a point source at P are reflected by a plane mirror and

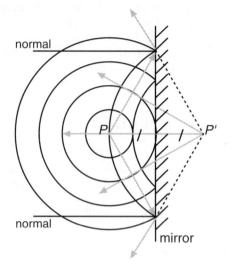

Figure 15.4 Waves from *P* are reflected in the mirror, and rebound as if from *P'*

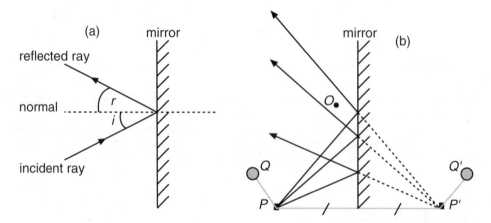

Figure 15.5 (a) Regular reflection at a plane mirror shows the angle of incidence, *i*, which equals the angle of reflection, *r*. In (b) three rays from *P* are reflected and spread out as if from *P'*. *P'* is the image of *P* in the mirror, and *Q'* is the image of *O*. The mirror line bisects both *PP'* and *QQ'*

each obeys the law of reflection. The three reflected rays appear to be spreading out from a point *P'* behind the mirror. Again *P'* is the image of *P* in the mirror, and the line joining *P* to *P'* is normal to the mirror.

In Fig. 15.6 straight waves in a parallel-sided beam approach a plane mirror. By the time the wave at *X* reaches *Y* after reflection at *Z*, it will have travelled the same distance as the wave at *X'* when it reaches *Y'* via the mirror at *Z'*. All the parts of the wave along *XX'* will also have travelled the same distance to form the straight wave *YY'*, where the angle of incidence equals the angle of reflection.

In Fig. 15.8 a concave mirror reflects light from a distant source and brings it to a focus in front of the mirror. The curved mirror changes the shape of the straight wave into a circular wave. Satellite dishes behave like this when they collect radio waves from a distant satellite and focus them onto an aerial in front of the reflector.

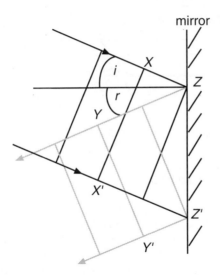

Figure 15.6 *Waves parallel to XX' are reflected at the mirror and become waves parallel to YY' heading in a new direction. The speed and wavelength of the waves are unchanged*

15.5 Lateral inversion

Any image in a plane mirror is as far behind the mirror as the object is in front of it. So in Fig. 15.5b we can tell that the image of the point Q is at Q'. Now consider what an observer at point O would see. Looking towards P and Q he or she would see Q to the right of P; but looking into the mirror, the image Q' is to the left of P'. So the image of an object between P and Q seen in the mirror would be reversed laterally left for right. You can check this by writing your name on paper and trying to read the writing by looking at its reflection in the mirror. The **laterally inverted** image of the writing appears the same size as the writing itself, just as PQ is as long as $P'Q'$ in Fig. 15.5b.

The image you see of yourself in a mirror is a laterally inverted you. It is not the picture others have of you.

.ɘnil ɘʜɈ ɿo bnɘ ɘʜɈ Ɉɒ ɿoɿɿim ɘʜɈ Ɉuq uoY

Figure 15.7

Self test

1 How could you use a second mirror to see yourself as others do?
2 Where should you place a mirror to read the writing in Fig. 15.7?

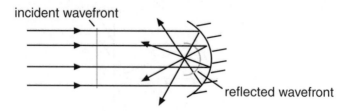

incident wavefront

reflected wavefront

Figure 15.8 A concave reflector brings rays to a focus where they cross. The mirror changes the shape of the wave

15.6 Virtual and real images

In Fig. 15.5b the reflected light waves do not pass through the mirror, but they appear to come from P'. The image at P' is not real but **virtual**. The curved reflector in Fig. 15.8 really does bring waves to a focus, and so the image of a distant object is a **real image**. A virtual image cannot be formed on a screen, and so the image on a cinema screen, a camera film, or in the eye has to be a real image where light is focused. A real image is one where the rays from an object come into focus, not just where they appear to come from.

15.7 Refraction

If you look down into a swimming pool, the water appears shallower than it really is, because light from a point on the bottom of the pool is refracted on crossing the surface from water into air. It emerges in the air in a new direction (see Fig. 15.9).

Fig. 15.10 shows what happens when rays cross a plane surface from air into glass. Parallel rays remain in a parallel sided beam, but they change direction on crossing the surface. The change of direction is explained by the **change of speed of light** when it crosses the boundary from air into glass in which it moves more slowly. There is also a

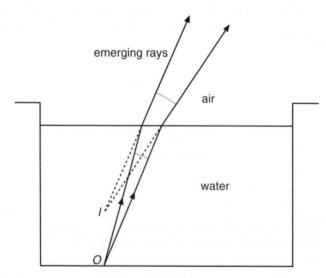

emerging rays

air

water

Figure 15.9 Light from O at the bottom of a pool is refracted at the water's surface. The emerging waves travel faster and diverge from a point I above the bottom, making the pool appear less deep

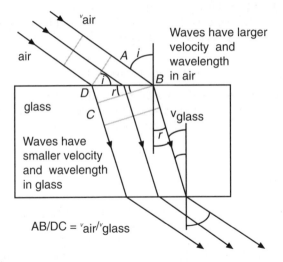

Figure 15.10 A parallel sided beam is refracted as it passes through glass

change of wavelength when light waves enter glass from air. A wave moving slower on entering glass starts to be caught up by the next wave which is still in air. Then the next wave enters the glass and the two waves move on at the new speed but with a new, smaller separation.

The speed of light depends on the substance it passes through, so the beam in Fig. 15.10 emerging from glass into air, speeds up and deviates away from the normal as it crosses the surface. On crossing from glass back into air, the light regains its original wavelength.

In Fig. 15.10 a wave in air reaches A and D at the same time. When the light travels along AB, it moves also along DC in the glass. So the ratio AB/DC is the ratio of distances gone in equal times. This can be written also as the ratio (speed of light in air)/(speed of light in glass). We can write an equation as:

$$\frac{AB}{DC} = \frac{AB \times BD}{BD \times DC} = v_{air}/v_{glass}$$

However AB/BD = sin i in Fig. 15.10, and DC/BD = sin r so the equation may be written as:

$$v_{air}/v_{glass} = \sin i/\sin r = n \tag{15.2}$$

The ratio of the two speeds is a constant independent of the direction of the light beam. The constant is called the **refractive index**, n, of the second substance in the first. This means that the ratio sin i/sin r must also be a constant for light of a given colour. Equation 15.2 expresses the law of refraction, which says that **when a ray of light of one colour crosses the boundary from one substance to another, the ratio sin i/sin r is a constant for the two substances and for light of this colour**. It is sometimes helpful to remember that $i/r = n$ is a good approximation, whenever the angles of incidence, i, and refraction, r, are both smaller than about 10°.

For light of a given colour we can apply the wave equation (12.1) to its waves in both substances:

$$v_{air} = f \times \lambda_{air}, \text{ and}$$
$$v_{glass} = f \times \lambda_{glass}.$$

No waves are gained or lost during refraction, so the number of waves per second moving into the glass is the same as the number leaving it. **The frequency remains the same throughout the process of refraction.** If we divide the two equations above we obtain the ratios:

$$v_{air}/v_{glass} = \lambda_{air}/\lambda_{glass} = \text{refractive index, } n, \text{ of glass in air} \qquad (15.3)$$

We see that the **ratio of the speeds is also the ratio of the wavelengths of the waves**. So when the speed is biggest, so also is the wavelength of the waves, and when the velocity is halved, the wavelength is halved, etc.

15.8 Dispersion

A ray of white light passing through a triangular glass prism is split into a band of colours, because the glass has a slightly different refractive index for each colour. Violet or blue light deviates more than red light, as the refractive index for blue light is slightly greater than for red (see Fig. 15.1). The angle between rays at opposite ends of the **spectrum of colours** is called the **dispersion**. In droplets of water this difference of refractive index for light of different colours gives rise to a rainbow when the sun shines while rain is falling in another part of the sky.

Suppose a second prism identical to the first were placed so as to receive light

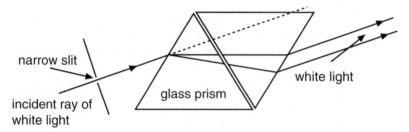

Figure 15.11 *The first prism splits white light into its separate colours, and the second prism combines these colours to make white light*

from the first prism in Fig. 15.11, causing it to deviate in the opposite direction. The two prisms side by side combine to form a parallel-sided glass block, and so the original ray of white light emerges from the second prism. We can see that white light is a combination of a range, or **spectrum of colours**.

15.9 Critical angle

A special property can arise when light is crossing the surface into a substance in which it travels faster. Fig. 15.12 shows that the light bends away from the normal as it travels from glass into air. At one angle of incidence called **the critical angle**, the angle of refraction is 90°. For any larger angle of incidence, the light does not emerge but is totally reflected. This phenomenon, called **total internal reflection**, is used in periscopes, prismatic binoculars, and fibre optics, which have applications in medicine and in telecommunications.

Fig. 15.13 shows the cross-section of an optical fibre consisting of an inner core

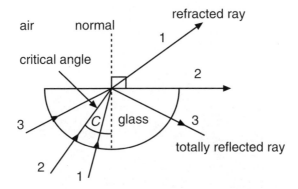

Figure 15.12 Rays 1 and 2 are refracted away from the normal on emerging from glass. The angle of refraction for ray 2 is 90°, and so the angle of incidence is the critical angle, C. Ray 3 makes a larger angle of incidence than C, and so it is totally reflected

surrounded by glass cladding with lower refractive index. Being only about 0.1 mm in overall diameter, the fibres are flexible enough to take gentle curves. A pair of optical glass fibres can carry 2000 simultaneous telephone calls without cross-talk. This means that each of the calls is kept separate by using separate frequencies of transmission along the same core. Fig. 15.14 shows fibre optics and metal cables capable of carrying comparable amounts of information.

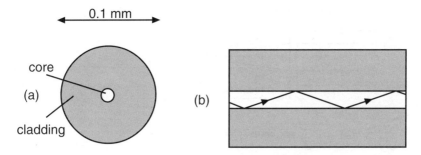

Figure 15.13 (a) Cross-section through an optical fibre shows the inner glass core surrounded by glass cladding of lower refractive index. In (b), light is totally internally reflected inside the core. The cladding has an outer protective coating to resist surface scratching and the risk of fracture

It is difficult to tap into a telephone fibre optic since the light in the core cannot be detected near the cable. There is a loss of signal strength as the pulses of light pass along the fibre optic, and booster units called **repeaters** are put into the line about every 25 km. Unlike microwave cross country links, fibre optic calls are not affected by electrical interference or by poor weather conditions, and so it is likely that fibre optics will be used increasingly as production costs come down.

15.10 Converging lenses

A converging lens is thicker in the centre than at the edges. Its curved surfaces change the shape of light wavefronts passing through the lens and can focus them to form images. The two types of image are illustrated in Fig. 15.15. They are **real** and **virtual images**. Real images like those on a cinema screen occur where light waves from an

Figure 15.14 Optical fibres on the left carry the same number of telephone calls as metal cables nearly ten times as thick on the right
(British Telecom)

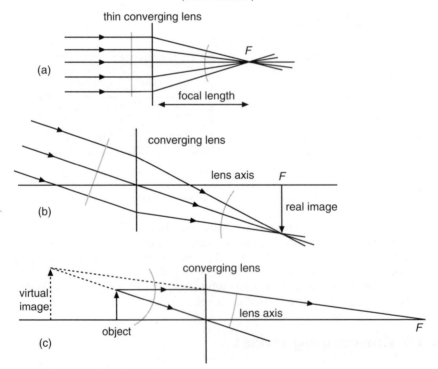

Figure 15.15 (a) A parallel-sided beam directed along the lens axis is refracted and passes through the focal point, *F*, of the lens. In (b) rays from a distant object form a real inverted image. In (c) rays from a nearby object emerge as if from the virtual magnified image. The shape of the wave is changed as the wave is delayed most on passing through the fatter central part of the lens

object are brought to a focus. A lens produces a real image if the object is more distant from the lens than a set distance called the **focal length of the lens**. This is the distance from the centre of the lens to the focal point, F, in Fig. 15.15a. The same lens can produce a virtual image if the object is brought closer to it than the focal length of the lens. Although the virtual image cannot be formed on a screen, it appears as a magnified version of the object on the same side of the lens as the object. A virtual image is formed by the eyepieces of both microscopes and telescopes. While the virtual image stands upright, the real image is upside down, or **inverted**. It is often said that a virtual image only appears to be there, while a real image is where the waves really do come into focus.

15.11 How to find the image position

There are two ways of finding the image position **provided you know the focal length of the lens and the object distance from it**. The first is by accurate scale drawing and measurement, and the second is by calculation.

Fig. 15.16 shows how to construct the paths of two special rays, out of all the very many rays that pass from the object through the lens. By finding where the two special rays come together, we find the plane where the image is formed. The construction of a ray diagram needs a sharp pencil and ruler and squared paper to work on.

The first ray to draw passes from a point at the tip of an arrow representing the object towards the centre of a thin lens. At its centre the lens behaves like a parallel-sided piece of window glass, and the ray passes through the lens undeflected.

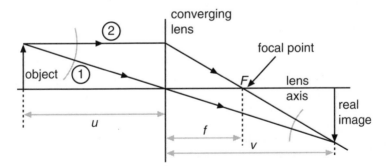

Figure 15.16(a) *Two special rays are plotted to locate the image. All the distances, u, v, and f are measured from the centre of the thin lens*

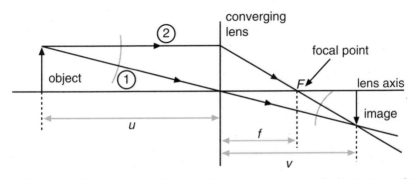

Figure 15.16(b) *When the object distance, u, from the lens increases, the image moves closer to the focal point of the lens, so v is less*

The second special ray to draw is the one from the tip of the arrow passing parallel to the lens axis. After passing through the lens this ray crosses its axis at the **focal point**, which is one focal length beyond the lens. The two special rays are continued until they cross where the image is formed.

To calculate where the image will be we use letters u, v, and f to represent the distances from the lens of the object, its image and the focal point respectively. The equation relating the distances is:

$$1/u + 1/v = 1/f \tag{15.3}$$

From this we can find that:

$$v = (u \times f)/(u - f) \tag{15.4}$$

A negative value for v indicates that the image is virtual.

15.12 Magnification

Making an object appear larger can enable more of its detail to be seen. The magnification produced by an optical system is the ratio:

$$\text{magnification} = \text{length of image/length of object} \tag{15.5}$$

As it is a ratio of two lengths, magnification has no units. If we measure the distances from a single lens or mirror, the magnification is the same ratio as:

$$\text{magnification} = \text{image distance/object distance} \tag{15.6}$$

Questions

1 When you look into a mirror, what two things about your image are exactly as others see you?
2 When you look into a mirror, what about your image is different from you as others see you?
3 If you step back two paces from a mirror,
   (a) in which direction does your image move?
   (b) how far does your image move from the mirror?
4 A beam of white light is reflected at each of two plane mirrors arranged at 90° to each other, Show that the emerging beam travels on a track parallel to the incident beam. (This principle is used for bicycle reflectors attached to the rear of bicycles).
5 Light travels at 3×10^8 m/s in air, and 2×10^8 m/s in glass. What is the refractive index of glass in air?
6 An object stands two focal lengths in front of a converging lens. Show that its image is two focal lengths beyond the lens, and find the magnification of the image.
7 A magnifying glass produces a virtual image with magnification $\times$ 2. Show that the image is one focal length in front of the lens.
8 In Fig. 15.17 numbers on a clock face are represented by dots. The hands indicate five minutes past eight o'clock. The clock face is reflected in the two mirrors A and B. What time is indicated by the reflection in (a) mirror A, (b) mirror B?

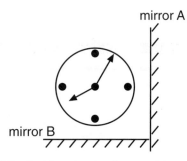

Figure 15.17 Clockface showing five past eight, and two mirrors

---

What have you learnt?

1　Can you show how waves are reflected at a plane surface?
2　Can you explain what happens to light waves as they cross a surface between substances?
3　Do you understand the term 'refractive index' of a substance?
4　Can you find where the image of an object is formed in a plane mirror or converging lens?
5　Can you distinguish between a real and a virtual image?
6　Can you explain what is meant by magnification?
7　Do you know what causes refraction of light?
8　Do you know the law of reflection?
9　Do you know how to split light into its colours?
10　Do you know what causes rainbows?
11　Can you explain the term 'critical angle'?
12　Do you know when total internal reflection occurs?
13　Can you name the parts of the eye, and the parts of a camera which perform similar functions?
14　Can you explain how a microscope eyepiece or magnifying glass works?

16 More about light waves

16.1 Introduction

We have seen how light waves are reflected and refracted in chapter 15. Now we shall consider the properties of diffraction, interference and polarisation.

16.2 Diffraction

While water waves flow round obstacles into the 'geometrical shadow' behind them, this is not so noticeable with light. However, the effect can be seen if you look through a very narrow slit at a narrow light source parallel to the slit, as in Fig. 16.1a. If the light is of just one colour, called **monochromatic light**, you see the light source as a wide bright line, on each side of which there are some less bright lines of the same colour, each half as wide as the central line.

The intensity of light varies across the field of view as in Fig. 16.1b. If the slit is adjustable, we find that the **pattern of light intensities becomes more spread out as the slit becomes narrower**. The intensities also reduce as the slit width is reduced, because less light can pass through a narrower slit than a wider one. The intensity pattern for a single slit shows that some of the light spreads out or **diffracts** on passing through the slit.

Diffraction of water waves is best seen when the obstacle or the gap through which the waves pass has a width close to their wavelength (see Fig.12.7). This suggests that

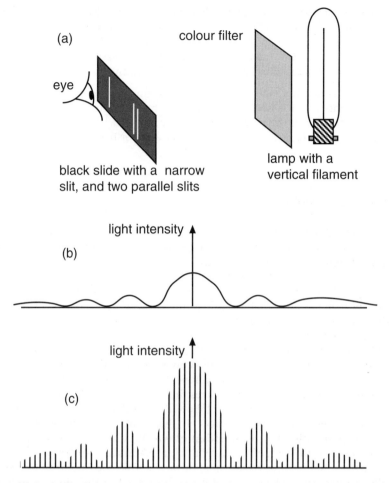

Figure 16.1 (a) The light source seen through a single narrow slit appears as a wide band of light whose intensity varies as in (b). (c) shows the interference pattern seen through the double slit

the wavelength of light must be very small, as the gap needed to demonstrate diffraction of light is very narrow. Now, if the wavelength of light is very small, we can understand why light does not diffract noticeably when it passes through windows, or even through the pupil of your eye. These are much wider than the wavelength of light. A slit only a fraction of a millimetre wide is best to show the diffraction of light.

16.3 How colour affects diffraction

The colour filter shown in Fig. 16.1a can be changed for a red filter at the top and a blue filter at the bottom of the filament bulb. When we look at the bulb through the same narrow slit, we see a red diffraction pattern above a blue diffraction pattern. **The lines in the red pattern are wider apart than the lines in the blue pattern**, though both are caused by light passing through the same slit. This shows that there is a difference between red and blue light. The red lines are wider because **red light has**

a longer wavelength than blue light. When we use other colour filters and examine the width of the lines in their diffraction patterns, we find their widths are bigger than the blue widths, but smaller than the red widths. From red to blue light there is a continuous change of wavelength, red having the largest wavelength and blue the smallest.

Single slit diffraction patterns can be recognised because:

(a) the central line is brighter, and is twice as wide as the lines on each side of it,
(b) the pattern of intensities is wider when the slit is made narrower,
(c) the width of the lines in the pattern illustrated in Fig. 16.1b depends on the colour of the light. For red light the lines are broader and further apart than for violet or blue light.

16.4 Young's slits

Light passing through two narrow parallel slits in front of the pupil of one eye produces a new pattern where the diffracted light from the two slits overlaps. This is a pattern of **equally spaced light and dark lines** as in Fig. 16.1c, whereas with one slit the central line was twice as wide as the others. The double slit pattern is a bit brighter than for one slit, as there is more light. There are many more lines than two, so this is not just the effect of light coming straight through the two slits. The spacing of the lines in these patterns is closer for blue light than for red, while the spacings for yellow and green light are somewhere in between.

We explain the bright lines in the pattern as the effect of light waves from the two slits reinforcing each other, and we call this **constructive interference**. Where light waves from two or more sources arrive **in phase** with one another, they reinforce each other to produce a visible line. In water waves this would occur when crests of two different waves arrived at a point together, followed a short while later by two troughs arriving together. The amplitude of waves at this point would be greater. This is according to the **principle of superposition of waves**, which says that the wave displacement at a point is the sum of the individual displacements arriving at the point at that instant. Waves that have travelled equal distances from their point source arrive in phase, as they would also do where one wave had travelled further than the other by a whole number of wavelengths.

The dark lines in the interference pattern are due to light waves from the two slits arriving exactly **out of phase** with one another, and according to the principle of superposition they cancel one another out. We call this **destructive interference**, and it happens where the distances travelled by the two waves from their source differ by an odd number of half wavelengths. We would explain this effect with water waves by saying that, if a wave crest reaches a point at the same time as a wave trough, they arrive out of phase and tend to cancel each other's effect.

Fig. 16.2 is a diagram of another experiment set up to produce an interference pattern with two slits, known as Young's slits. Here the lamp is close to the slits, and the fringe pattern is observed in a plane at distance D beyond the slits. The figure is not to scale, as the separation of slits, s, is less than 1 mm, and the distance OP on the screen is very much less than D. The path S_1P and S_2P to the screen differ in length by almost exactly S_2T, and the small angles, α and β, are almost identical. Angle β in the figure is TS_1S_2.

Suppose there is a bright fringe at P. This means constructive interference occurs, as S_2T is a whole number, n, of wavelengths, i.e. $n \times \lambda$. For darkness at P, then S_2T would be an odd number of half wavelengths, making the two sets of waves arrive out of phase.

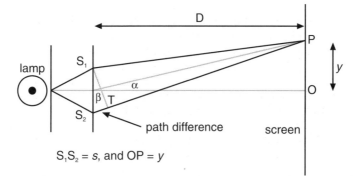

lamp

S_1

α

β T

S_2

path difference

D

P

y

O

screen

$S_1S_2 = s$, and OP = y

Figure 16.2 *Light of one colour emitted by the lamp travels equal distances via one slit to slits S_1 and S_2. Diffracted light from these slits reaches a screen where equally-spaced light and dark fringes appear, due to interference of waves from the two slits, S_1 and S_2*

For small angles we can equate tan α and sin β. So:

$$y/D = n \times \lambda/s \text{ where } s = S_1S_2 \qquad (16.1)$$

where n = 0, 1, 2, 3, ... etc. The distance between one fringe and the next is the change in y when n changes by 1. This is $(D \times \lambda)/s$; it shows that the **larger fringe spacing for red light** than for blue occurs because **red light has a longer wavelength than blue**. We can use the experiment to find the wavelengths of light.

The condition for darkness at P is that S_2T is an odd number of half wavelengths, i.e.

$$y/D = (2n + 1)\lambda/2s \qquad (16.2)$$

16.5 Diffraction gratings

A diffraction grating has a series of equally-spaced, very narrow parallel lines, scratched close together on a glass surface. Light of one colour passing through the gaps between the scratches is diffracted and the resulting bright fringes occur where waves from many slits interfere constructively. For the least deviated light, or the **first order**, the extra distance travelled by light passing through one slit as against the next slit on the grating, is one wavelength as in Fig. 16.3. For the more deviated light in the **second order** the extra path length is two wavelengths, etc. The spacing between the lines on the grating is recorded when it is made, and with this data it is possible to find the wavelength of the light.

Using Fig. 16.3 we may write that for the nth order bright line:

$$\sin \theta = n \times \lambda/s \qquad (16.3)$$

Fig. 16.4 illustrates a laboratory experiment for finding

the wavelength of the yellow light from a sodium vapour lamp. The grating is mounted near one end of a metre rule, the other end of which is pointed at the distant light source. The observer sees a first order diffracted image of the lamp at an angle θ to the metre rule. A pin is lined up with this image supported on a half metre rule clamped at right angles to the metre rule, and the distances x and y are recorded.

We calculate the angle the light has been diffracted in the first order using the equation:

$$\tan \theta = y/x \tag{16.4}$$

When θ is known, we can calculate the wavelength, λ, using the equation (16.3) where $n = 1$. So:

$$\lambda = s \times \sin \theta \tag{16.5}$$

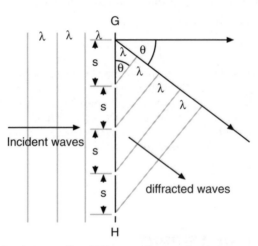

Figure 16.3 The grating between G and H has equally-spaced gaps for waves to pass through. The diffracted waves reinforce one another only when their path lengths differ by a whole number of wavelengths

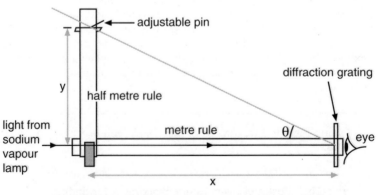

Figure 16.4 Apparatus for finding the wavelength of light

The diffraction grating with as many as 6000 lines per cm can be used to obtain accurate values for the wavelengths of light emitted by gases in low pressure electrical discharge tubes. The wavelengths of light in air range from about 420×10^{-9} m for violet light, to about 750×10^{-9} m for red light. These may be written 420 nanometres (nm), and 750 nm. One nanometre is 10^{-9} m.

Example
If $\theta = 20°$ for deviated light in the first order spectrum with a grating having 6000 lines per cm, find the wavelength of the light.

Answer
Use equation 16.5 above with $s = 1/6000$ cm.

ie. $\lambda = (\sin 20°/6000)$ cm $= 5.70 \times 10^{-5}$ cm
$= 570 \times 10^{-9}$ m $= 570$ nm.

An easy way to demonstrate diffraction using a reflection grating is to study the reflection of light from the surface of a compact disc. If you stand beneath a filament light holding the CD so that it reflects light up into your eyes, you see the image of the light source. Then as you tilt the CD, the light reflected from its many closely spaced lines will be seen, first the blue end of the visible spectrum and then the other colours up to red. If you tilt the CD further, you see the same sequence of colours again in the second order spectrum. In the first order spectrum the light reflected at one line on the CD has travelled one wavelength further than the light from the neighbouring line. In the second order, the extra path difference for light reflected at neighbouring lines is two wavelengths. The separation of the colours is caused because they have different wavelengths.

16.6 Polarised light

The displacements in a transverse wave are perpendicular to the direction of wave motion. Waves on a rope are like this. But if the rope passes through a narrow slit, only displacements parallel to the slit are transmitted through it. Wave displacements perpendicular to the slit are either reflected or absorbed when they reach the slit. So if the slit is vertical, only vertical displacements pass through it, and the waves are then said to be vertically polarised.

Light can also be polarised on passing through polaroid material which selectively absorbs the wave displacements in one direction, while it transmits those in the perpendicular direction. Two layers of polaroid material may be held so that they transmit polarised light (see Fig. 16.5). However the intensity of the light is reduced when one of the polaroids is rotated in its own plane. After a rotation of 90° as in Fig. 16.6, the polaroids are 'crossed', and virtually no light emerges from the second layer.

Since longitudinal waves cannot be polarised, and light can be, this is evidence that light is a transverse wave motion.

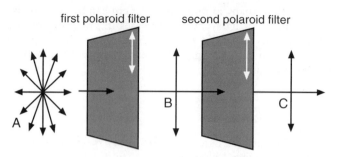

Figure 16.5 An unpolarised transverse wave at A has displacements in a plane at 90° to its direction of travel. At B the horizontal wave displacements have been filtered out. The wave is polarised vertically. The second polaroid filter absorbs only a fraction of the light it receives, and so polarised light reaches C

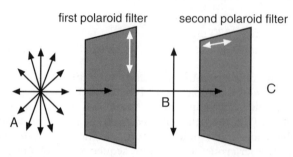

Figure 16.6 Light waves at B are vertically polarised. The second polaroid filter has been rotated so as to transmit only horizontal displacements in the waves, and so no light reaches C

16.7 Electromagnetic spectrum of waves

The colour of light is related to its frequency and wavelength. Beyond the ends of the visible spectrum other waves may be generated although we cannot see them. They may be detected by photographic or other means, and they have some of the same properties as visible light. Their common properties are:

(a) they all travel at the speed of light through an evacuated space ($c = 3 \times 10^8$ m/s)
(b) they are all transverse waves, each with its own frequency and wavelength
(c) they are all generated by the movement of electric charge.

Among the family of waves which make up **the electromagnetic spectrum** are:

γ rays, X-rays, UV rays, visible light, IR rays, microwaves, radar, TV and radio waves.

These are listed in order of increasing wavelength and decreasing frequency. The largest wavelength radio waves have the lowest frequency, and vice versa.

1 Light of wavelength 600 nm in air crosses a surface into water of refractive index 4/3.
   (a) What happens to the speed of the light?
   (b) What happens to the wavelength of the light?
   (c) What happens to the frequency of the waves?
2 Find the frequency of light of wavelength 600 nm in air where it travels at 3×10^8 m/s.
3 (a) If a grating has 7500 evenly spaced lines per cm, what is the line separation in metres?
   (b) A parallel-sided light beam meets this grating perpendicularly, and it is deviated 30° in the first order. Find the wavelength of the light.
4 How could you tell if you were looking at a single slit diffraction pattern or a double slit interference pattern?
5 How would the spacing of fringes in a double slit interference pattern be changed if:
   (a) the wavelength of the light were increased?
   (b) the slits were spaced closer together?
   (c) the pattern was studied at a greater distance from the slits? (see Equation 16.1)
6 If a grating has 6000 lines per cm, find the first order deviation of diffracted light of wavelength 650 nm.
7 Why can 300 m wavelength radio waves diffract round hills into valleys, while television signals of wavelength less than 1 m are best received by aerials in sight of the transmitter aerial?
8 (a) Name a property of light waves that sound waves cannot exhibit.
   (b) Explain why sound waves are unable to show this effect.

What have you learnt?

1 Can you recognise a single slit diffraction pattern for monochromatic light?
2 Do you know how constructive interference occurs when light reaches a point by two routes?
3 Do you know what an interference pattern obtained with Young's slits looks like?
4 Do you know what a diffraction grating is?
5 Can you explain why there may be first order and second order diffraction patterns when light passes through a diffraction grating?
6 Can you use the grating theory to find the wavelength of diffracted light?
7 Do you know that only transverse waves can be polarised?
8 Do you know the family of waves called the electromagnetic spectrum?
9 Do you know why colours appear in light reflected from the surface of a CD?

Making use of waves

After completing this chapter you should
- understand the functions of parts of the eye
- understand the eye's means of focusing
- be able to explain long and short sight, and ways of correcting for them
- know the parts of a camera and how they function
- understand how optical and radio telescopes are helpful to astronomers
- understand some uses of radio waves, radar, microwaves and infrared radiation
- know that there are different kinds of earthquake waves
- understand how ultrasonic waves can be used.

17.1 Introduction

Having seen the properties of waves, we consider their various uses.

17.2 The human eye

Fig. 15.1a showed a simplified diagram of the human eye, of which two important parts are the lens and the retina. From chapter 15 we know that light which passes through the eye's converging lens is refracted to form a real inverted image in focus on the retina; and as the retina is sensitive to light, it sends signals to the brain which convey the shape, size, colour and position of the object in view. The image is completely inverted, but the brain gets used to interpreting this as being caused by an object which is 'the right way up and the right way round'.

17.3 Seeing things clearly at different distances

In chapter 15 we found that as an object is moved further from a converging lens, its image is formed closer to the lens, and vice versa.

Suppose that the lens is in the eye, and that the image of a distant object is in

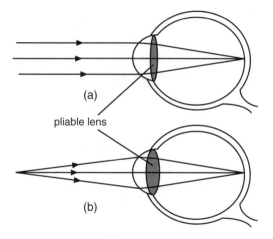

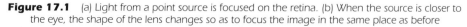

Figure 17.1   (a) Light from a point source is focused on the retina. (b) When the source is closer to the eye, the shape of the lens changes so as to focus the image in the same place as before

focus on the retina. If the object is now brought closer to the lens, its image position is behind the retina, and so the image on the retina is out of focus.

The eye adjusts automatically for this change by changing the shape of its lens, so that the image distance remains unchanged for the shorter object distance (see Fig. 17.1). Thus the image is still in focus on the eye's retina. For a normal-sighted person the lens shape can adjust so that an object from 'infinity' down to about 30 cm from the eye can produce an image in focus on the retina.

17.4 Short sight

A short sighted person can focus on things nearby, but not on things a long way off. The eye lens bends light too much, or the retina is too far from the lens, or both. So light from a distant object forms an image in front of the retina.

The cure for this is to wear glasses in which each lens diverges the light before it enters the eye (see Fig. 17.2), so that it can be focused exactly onto the retina.

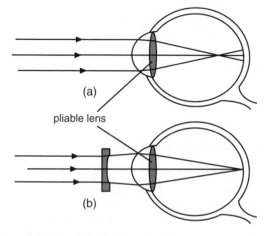

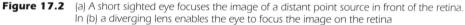

Figure 17.2   (a) A short sighted eye focuses the image of a distant point source in front of the retina. In (b) a diverging lens enables the eye to focus the image on the retina

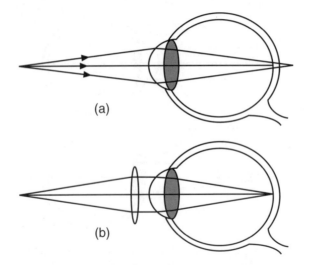

Figure 17.3 (a) A long sighted eye focuses light from a close object in a plane behind the retina. In (b) a converging lens enables the eye to focus the image correctly on the retina

17.5 Long sight

A long sighted person can focus on things a long way off, but not on things nearby. The eye lens does not bend light enough, or the retina is too close to the lens, or both. The result is that objects at a convenient reading distance form images behind the retina.

The cure for long sight is to wear glasses with weak converging lenses, which assist the eye's lenses in forming real images in focus on the retina (Fig.17.3).

17.6 Old age sight or presbyopia

As people get older their eye lenses become less pliable and cannot change shape so much. The result is often similar to long sightedness. That is to say, that while they can see distant objects clearly, they cannot easily see nearby objects because their images would be in focus behind the retina.

The condition is corrected by spectacles with converging lenses which assist the eyes' non-pliable lenses, so that clear images of close objects may be formed on the retina (see Fig. 17.3).

17.7 Seeing coloured objects

The colours we see can all be produced on a white screen by superimposing beams of three primary colours, red, green and blue with various relative intensities (see Fig. 17.4). The retina contains three types of receptor, sensitive respectively to each of the primary colours. So if, for example, yellow light falls on the retina, then both the red sensitive and green sensitive receptors are stimulated. The dual signal sent to the brain causes the person to 'see' yellow.

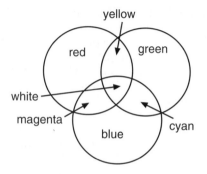

Figure 17.4 The addition of primary colours red, green and blue on a white screen. Yellow, cyan and magenta are the secondary colours

---

Self test ───

Which two light receptors would be stimulated when we see a magenta cloth?

17.8 Colour blindness

This is a condition which is found in a small proportion of the male population and very few females. It is usually a condition inherited from the mother, though she is unlikely to be colour blind. If her father was colour blind then she may transmit this to her sons. Colour blindness means that one may not be able to distinguish some colours from one another, even though objects are seen clearly in focus.

17.9 The slide projector

The principal parts of the slide projector are illustrated in Fig. 17.5. Light from a bright filament lamp illuminates the slide to be viewed. Light emitted away from the

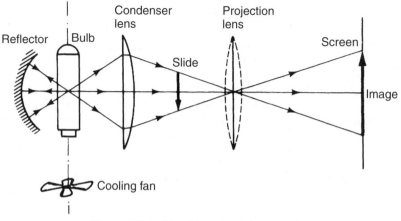

Figure 17.5 The slide projector (not to scale)

slide is reflected at a concave reflector, so that the amount of light reaching the condenser lens is increased, as is the illumination of the slide.

The illuminated slide is the object for the projection lens, which forms a real, magnified and inverted image of the slide on a distant screen. If the final image is to be the correct way up, the slide must be loaded on the slide carrier in an inverted position. The screen is usually several metres from the projection lens, whose focal length is typically about 10 cm. This means that the slide must be slightly more than 10 cm from the projection lens.

Projector bulbs produce a good deal of heat as well as light. Two features are added to the projector to allow for this. First a fan is fitted to cool the bulb's surface by forced air convection; and secondly there is a special glass filter (not shown in the figure), between the bulb and the slide to protect the slide from damage due to infra-red radiation.

17.10 The camera

The most common type of camera (Fig. 17.6) consists of a light-tight box containing one opening in which there is a converging lens. The lens produces real images in focus on the film at the other end of the box.

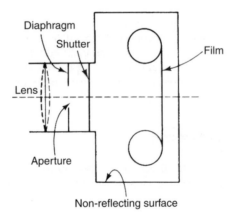

Figure 17.6 A camera

Just behind the lens is a light-proof shutter which stops light getting to the film until required. When a photograph is to be taken, the shutter is drawn momentarily away from the lens, letting light through onto the film for a precise length of time.

The size of the hole through which light passes to the film can be adjusted. The hole, or aperture, occurs in a diaphragm placed behind the lens near the shutter.

Objects at different distances from the lens give rise to images at different distances beyond it. If the images are to be in focus on the film, the lens must be moved until the distance from the lens to the film matches the image distance for each particular object.

The image on the film is, of course, inverted like the image formed on the retina of the human eye.

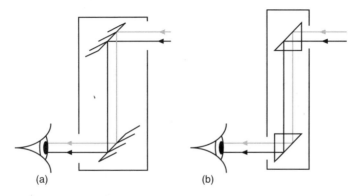

Figure 17.7 (a) shows a periscope made with two plane mirrors. In (b), the periscope is made with two glass prisms. They have no silvered surfaces but use total internal reflection

17.11 The periscope

Fig. 17.7a shows the use of two plane mirrors to produce a periscope. The first mirror produces a lateral inversion of the original object. The second produces a further lateral inversion, which cancels out the first, so that the final image is the same way up and the same way round as the original object.

Normal silvering on a mirror reflects only about 70% of the light incident on it, the rest being absorbed by the silver. Furthermore the silver gradually tarnishes. A prism, however, which does not tarnish, reflects almost all the light incident on it, provided the conditions are met for total internal reflection to occur (see Section 15.9). So a periscope using prisms is a practical proposition for a submarine; one using silvered mirrors is not.

Fig. 17.7b shows a periscope suitable for a submarine. The mirrors are replaced by right-angled prisms. Light waves incident normally on the face of the prism as shown, meet the internal face at an angle of incidence of 45°. This is greater than the critical angle for glass, and so the light is totally internally reflected. The same thing happens at the second prism, which also behaves as a mirror.

17.12 Optical astronomy

Many large telescopes used in astronomy these days involve the property of a concave mirror to converge electromagnetic waves (Fig. 17.8).

Light from a distant star or galaxy falls onto the mirror, which in the telescope at Mount Palomar, California has a diameter of 500 cm. After reflection it forms an image which can be photographed or viewed directly through an eyepiece lens. The bigger the diameter of the main mirror, the better is the telescope at revealing detail. However, there is a limit to the detail that can be seen because the light has to pass through an obscuring layer of atmosphere at the end of its journey from a distant star.

The Hubble Space Telescope, with its mirror 240 cm in diameter, orbits the Earth at a height of about 600 km (see Fig. 17.9). This is above the obscuring and distorting effects of the atmosphere. So, although the mirror diameter is only about half that of the Mt Palomar mirror, the pictures it produces are much clearer.

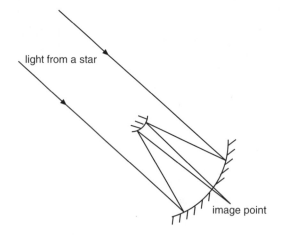

Figure 17.8 This reflecting telescope has a mirror with a large diameter causing light from a star to converge. A second mirror brings the image into focus behind a central hole in the large mirror. The image in viewed through a lens eyepiece (not shown)

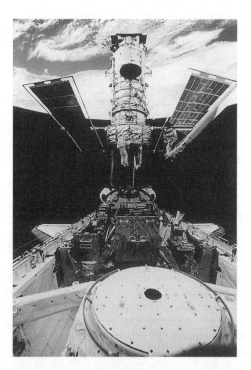

Figure 17.9 Astronauts service the orbiting Hubble Space Telescope. You can see the solar panels on each side of the telescope, used to power its instruments. The photograph was taken from the open bay of the Space Shuttle
(NASA)

The Hubble Space Telescope has an optical system like that of Fig. 17.8. The outer casing of the whole package is about as big as a single decker bus, with a length of 13 m, and diameter 4.5 m. The unfolded solar panels add to these dimensions.

Radiations from ultraviolet, through visible, to infra-red are focused behind the main mirror in a bay where they are monitored and analysed. It is reckoned that this system could detect light from a firefly at 16 000 km. The two photographs in Fig. 17.10 illustrate the advantage of

Figure 17.10 The best Earth-bound telescope, even if computer-aided, produces limited detail; but (b) the Hubble Space Telescope shows much greater detail of the same part of the sky (NASA)

placing the telescope in orbit above the Earth's atmosphere. Fig. 17.10a shows the best that can be done when looking at a small faint region of the sky with a land-based telescope. Fig. 17.10b shows the same part of the sky seen with the Hubble Space Telescope. The improved picture quality enables astronomers to collect data about more distant and fainter objects than ever before.

Figure 17.11   The Lovell Radio Telescope at the Nuffield Radio Astronomy Research Laboratories (The University of Manchester)

17.13 Radio astronomy

Objects in space do not just emit visible light. Radiation which they emit in other parts of the electromagnetic spectrum may also be received and analysed.

The Lovell Radio Telescope at Jodrell Bank, Cheshire, England, (Fig. 17.11) uses a concave reflector of diameter 76 m to bring to a focus radio waves moving along the reflector's axis. At the focus is a detector which registers the strength of the incoming signals. The detector is mounted on top of the tower you can see in the photograph. This telescope is a large scale version of a TV dish aerial which receives signals from an orbiting satellite.

> One advantage of a wide reflecting 'dish' is that more energy is collected from each radio source in space, and so fainter radio stars can be detected. A second advantage of a wider 'aperture' telescope is that it can show up finer detail. This sometimes reveals that one apparent radio source is in fact at least two sources, detected in almost the same direction. Radio astronomers have been able to link several widely separated radio telescopes so that their combined signals enable even finer detail to be observed. The telescopes have to be directed to receive their signals from the same part of the sky, and the signals are combined using overland cables, or by means of

microwave relay stations linked to Jodrell Bank. Linked in this way, seven radio telescopes can act as one whose receiving dish is 217 km wide (Jodrell Bank to Cambridge). The detail revealed is then similar to that shown by the Hubble telescope using shorter wavelength visible light. In theory the performance of a telescope in showing fine detail improves as its ratio (width/wavelength) increases.

17.14 Radio and TV waves

Radio waves and television waves are two parts of the whole group of waves which make up the electromagnetic spectrum mentioned in chapter 16.

The difference between the categories is their wavelength. Radio waves have a range of wavelengths from about 10^5 m to 10 m, and TV waves a range from about 10 m to 10^{-2} m. All of these waves can be used to send information from transmitter to receiver (radio or television). The information can consist of television pictures, or sound like speech or music, or both pictures and sound.

Being electromagnetic waves, each has both electrical and magnetic properties. For the shorter wavelength it is easier to 'pick up' the signals by detecting the electrical effect, using metal wires in TV aerials. Small currents are generated by the passing waves. At longer wavelengths it is easier to detect the effect of the waves passing a magnetic 'ferrite' rod aerial inside the radio receiver.

17.15 Colour TV

Colour television pictures are made up of combinations of three primary colours emitted from a series of minute fluorescent dots. The dots emit either red, green or blue light when struck by a beam of fast-moving electrons. There are three separate beams of electrons inside the tube and each generates a picture in one of the primary colours. Just in front of the TV screen is a metal plate called a shadow mask with a series of holes in it. The shadow mask ensures that each beam of electrons reaches only one set of dots to generate one primary colour (see Fig. 17.12).

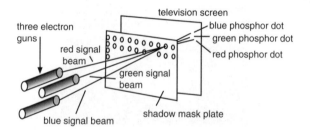

Figure 17.12 A colour television tube, having three electron beams controlled by the received TV signals, produces a composite picture in red, green and blue dots

As in the cathode ray oscilloscope (chapter 23), the television set contains a time-base circuit which makes the three beams sweep across the screen at high speed. In 1/25 second they sweep across 625 times and move from top to bottom of the screen. This is so rapid that the eye sees a continuous picture.

The intensity of each colour on a part of the screen depends on the number of electrons in the beam striking it. The energy in each electron stream is adjusted by the incoming TV waves, which are detected by the aerial and produce electronic signals to the TV tube.

17.16 Radar

The word 'radar' comes from the label '**ra**dio **d**etection **a**nd **r**anging', used to describe the original system. Modern radar systems can do more than detect and establish distance, but the same term is still used.

Radar systems transmit electromagnetic energy with a wavelength in the region between about 1 mm and 1 m. The waves are transmitted as a series of pulses each lasting about 5 microseconds.

The direction of transmission of the waves is determined by a rotating aerial. You may see these aerials at airports or on ships. If a pulse of energy encounters a reflective object, e.g. an aircraft, cliff or ship, then some energy is reflected back and may be detected by the same aerial acting as a receiver (Fig. 17.13). From the time delay between the production of the pulse and its reception after reflection, it is possible to find out the distance to the reflecting object. The position of the rotating transmitter and receiver aerial at that instant indicates the direction of the object.

Figure 17.13 A rotating radar transmitting and receiving aerial at Heathrow Airport, London (Civil Aviation Authority)

17.17 Microwave ovens

Microwaves are electromagnetic waves with wavelengths between about 1 mm and 50 cm. Microwaves of a wavelength of about 12 cm, with a corresponding frequency of 2450 MHz, have a domestic use in microwave ovens.

The microwaves are absorbed by water, fats, sugars and certain other molecules, but not by glass or china. So if some food in a glass container is placed in a stream of microwaves, the food will absorb the radiation and warm up, while the radiation passes straight through the container.

The oven is usually a metal box, apart from its door, and metals reflect microwaves. There is a possibility of uneven heating in the oven, but this is overcome by moving the food continually by standing it on a rotating platform.

17.18 Earthquake waves

Earthquakes create waves which are not electromagnetic. An earthquake occurs when there is a sudden slip of one part of the Earth's crust against another. The effect is like shaking a large block of jelly. Waves are sent out through the Earth from the point of slip. They can be of two kinds, transverse or longitudinal, and they travel through the Earth at different speeds. Near the Earth's surface longitudinal waves (called P waves) travel at 5.57 km/s, and transverse waves (called S waves) travel at 3.56 km/s. This sometimes enables appropriate warnings to be given.

For example, when an earthquake occurs, both transverse and longitudinal waves are sent out from the earthquake centre simultaneously. The faster-travelling longitudinal waves arrive at any recording station before the slower transverse waves. From the time delay between the arrival of the two sets of waves, it is possible to calculate the distance of the earthquake's centre. If two or three recording stations do this calculation, the exact origin of the earthquake can be established.

The Pacific Ocean is ringed by earthquake-prone regions. If an earthquake occurs close to or underneath the sea, waves will be sent out through the Earth as explained above. In addition, waves may be sent out across the water's surface, with small amplitude in deep water, but with damagingly large amplitude in the shallower water of harbours, docks and beaches. These water waves travel at about 800 km/h, and cross the Pacific Ocean in about 5 h. In contrast the waves which travel through the Earth take 5–6 minutes to travel the same distance.

If the earthquake recording stations are continuously alert, they can determine where an earthquake originated, and when. Then they can issue a warning of an impending earthquake water wave, or 'tsunami', so that those living near the water's edge, or with boats in harbours, can take precautions.

17.19 Using ultrasound

Ultrasound refers to sound waves of such a high frequency (above about 20 kHz), that it is inaudible to humans. Even so, some ultrasound can be heard by animals like dogs and bats. Audible sound has frequencies in the range from about 40 Hz to 20 kHz. It spreads out easily in all directions from its source; but ultrasound with its higher fre-

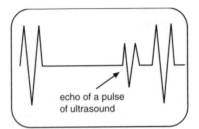

echo of a pulse
of ultrasound

Figure 17.14 The CRO trace shows an echo from an object beneath a ship arriving between two outgoing, larger amplitude pulses of ultrasound

quency and shorter wavelength can be confined to a more narrow beam, because diffraction occurs much less than for audible sounds.

Ultrasound undergoes reflection and refraction at a boundary between two substances, and this forms the basis for its use in various fields. Three examples are given below.

(a) *Echo sounding*

Suppose a boat sends out short bursts of ultrasound waves at 40 kHz. An oscilloscope screen on the boat can show the effect of two such short bursts and, in between, an echo received on the boat from the sea bed (Fig. 17.14).

The horizontal scan on the oscilloscope screen is used to indicate the time delay between producing a burst of ultrasound and receiving its echo. The depth of water is calculated from the delay time and the speed of sound in water, i.e. total distance gone = 2 × depth = speed of sound × delay time.

Self test

The time delay between producing a burst of ultrasound beneath a ship and receiving its echo is 16.8 ms. Show that the sea bed is 12.6 m below the sound emitter.

(Hint: use speed = 2 × depth/time lapse. Speed of sound in water = 1500 m/s.)

Sound echoes in nature
Dolphins at sea and bats in flight both use sound pulses for communicating. They also display range-finding skills by using the delay times of echoes to indicate their distances from reflecting objects around them.

Bats, by changing the wavelength and pitch of sounds they emit, may find the size and shape of a reflector as well as its direction and distance away. The Noctule bat, whose wingspan is 32 to 40 cm, emits two types of ultrasound when foraging above the tree tops of a forest. One pulse is of fairly steady frequency in the range 18 to 25 kHz, and this alternates with a shorter pulse of changing frequency in the range from about 25 to

45 kHz. The second of these pulses has shorter wavelengths, which could reveal more detail about a small target's size.

Ultrasound reflections from insects in flight can reveal their velocities. An insect flying towards the bat reflects more waves each second than if it were at rest. So the echo reaching the bat has a frequency slightly higher than the bat emitted. Similarly, an insect flying away from the bat reflects fewer waves each second, which lowers the frequency of the echo. The amount by which the frequency of the echo is changed indicates the speed of the insect moving towards or away from the bat.

To defend themselves against being eaten by bats, some moths settle under leaves as soon as they detect pulses of ultrasound. Their echoes then have the same frequency as those from the tree, and so the moths become harder to detect.

(b) *Medical scanning*

Reflections or echoes of higher frequency ultrasound from tissue boundaries within a human body enable medical experts to build up an ultrasound image of parts of the body, without having to cut into it for a direct look. The hazards of ultrasound are less than those of using X-rays, and ultrasound of frequency above 1 MHz is a useful medical tool.

For example, ultrasound images of an unborn baby can be obtained with no discomfort to the mother or baby. A very narrow beam of very high frequency ultrasound is generated in a source held close to the mother's womb. An oscilloscope screen registers the echoes which are detected by reflections from the boundary between muscle and bone, or between lung and heart tissue. In Fig. 17.15, the direction OA on the

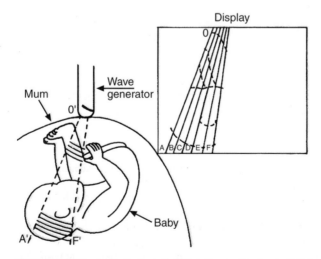

Figure 17.15 An illustration of ultrasonic scanning in medicine

screen indicates the same direction O'A' through the mother's body, and any echo from the mother's body along O'A' shows up as a bright spot on the oscilloscope screen. As the signal generator sweeps through O'A', towards O'F', a picture is built up of the inside of the mother's womb, including the unborn baby. Similarly a picture may be obtained of the organs inside any part of a patient's body.

In order to get a sharp picture in medical scans, the ultrasound frequency must be in the range from 1 to 20 MHz, or else diffraction effects make the image too blurred to be useful.

(c) *Ultrasonic cleaning*

Large amplitude ultrasonic waves can be generated and sent through a liquid. They can be used to clean metal surfaces immersed in the liquid. The ultrasonic waves produce forces that can break off small particles and contaminating material from the metal surface.

Questions

1   Which of these words describes the image formed in the eye on the retina: (a) real, (b) virtual, (c) upright, (d) inverted, (e) magnified, (f) diminished?
2   List the three primary colours, and the three secondary colours, formed by the addition of pairs of primary colours on a white screen.
3   How is a slide in a slide projector protected from overheating?
4   (a) If you are long sighted, what objects might you have difficulty in seeing clearly?
    (b) What sort of spectacle lenses might assist your eyes to see more clearly?
    (c) When would you need to wear these glasses?
5   What features in a slide projector increase the brightness of the projected image?
6   Which part of the slide projector focuses the projected image?
7   Explain with the aid of a diagram how glass prisms are used in a periscope.
8   Why is ultrasound and not audible sound used to scan organs in the body?
9   If sound travels at 320 m/s in air, find the wavelength of ultrasound of frequency 3.2 MHz.
10 What is the typical wavelength of microwaves used for cooking?
11 What sort of waves are created by earthquakes centred (a) on land, and (b) under the oceans?
12 A camera is set up to photograph in close-up. How must the lens be adjusted to photograph a more distant object?
13 Name two types of wave which are not electromagnetic.

What have you learnt?

1   Do you know how the eye (a) adjusts itself for variable light intensities, (b) focuses images?
2   Do you know how glasses can assist those with (a) short sight, and (b) long sight?
3   Can you draw a diagram of a camera and explain how it works?
4   Can you explain how a slide projector works?
5   Can you draw a diagram and explain how a periscope works?

6   Can you explain the advantage of using a large diameter mirror in a reflecting telescope.
7   Do you know how radar is used in navigation aids?
8   Do you know how a colour TV screen works?
9   Do you know how ultrasound can be used?
10 Do you know that earthquakes may set up both transverse and longitudinal waves?

18 Introduction to electricity

Objectives

After reading this chapter you should
- know that an electric current is a flow of charges in conductors
- know that a cell or battery can drive a current in a loop of conductors called a circuit
- know the units of charge and current
- explain the sizes of currents when conductors are connected in series or parallel
- know that the driving force of a cell (its e.m.f.) is measured in volts
- know how ammeters and voltmeters are connected in circuits
- understand that volts are joules of energy transferred per coulomb of charge
- know the symbols used in circuit diagrams.

18.1 Static electric charges

A plastic comb, after being rubbed against a fabric, may be used to pick up a small piece of paper nearby. The effect is caused by electric attraction between charged particles. In the rubbing action, the comb's surface gains from the fabric some negative charges, which can then attract charges in the paper. The comb's surface becomes negatively charged.

Electric attraction can also be seen when the teeth of the charged comb are held close to water trickling from a tap. The falling water is pulled sideways towards the comb (Fig. 18.1).

Uncharged matter contains equal amounts of positive and negative charge. Experiments show that **like charges repel each other, while unlike charges attract each other. The force of attraction or repulsion is greater when the charges are closer together**.

Water molecules, like those in paper, have positive and negative charges. In molecules of water the two types of charge are slightly separated, so that a molecule can rotate its positive centre towards the negative charge on the comb, leaving the negative centre of charge slightly further away. The comb now attracts the positive charges towards it and repels the negative charges. The resultant force is a net attraction of the water towards the comb, because the positive charges in the water molecules are centred slightly closer to the comb than their corresponding negative charges.

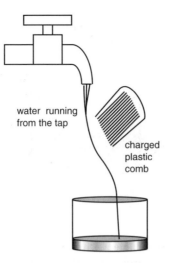

Figure 18.1 *The water is drawn towards the teeth of the comb*

In a similar way negative charges on the rubbed comb cause a rearrangement of molecules in the paper, and the paper is attracted towards the comb.

18.2 Insulators and conductors

Insulators like glass, rubber and plastics may hold surface electric charge without enabling the charge to flow through them. There are only two types of electric charge, and we call them 'positive' and 'negative' charges. Positive charges are carried by minute particles called **protons**, which are part of the central nuclei of atoms.

Electrons carry negative charge, each as big as a proton's positive charge. Electrons are particles of even smaller mass than protons, and they are found on the outsides of atoms. Because the negative charges are on the outsides of atoms, they are more likely to be shared between atoms, removed from atoms, or even added to atoms or molecules. Chemists study these changes.

Electrical conductors include all metals. A small fraction of the electrons in metals are freed from their parent atoms, and free electrons can flow through the material; **so a current in a wire consists of the flow of a huge number of these tiny negative charges**. Copper and silver are very good conductors.

Semiconductors like germanium and silicon also conduct electricity. They conduct less well than metals, as they possess fewer moveable 'conduction' electrons.

Both gases and liquids may conduct electricity, provided they contain **charged molecules called ions**. In liquids the current can be carried by positive ions flowing in one direction and negative ions flowing in the opposite direction (see Fig. 18.2). The flow of charge in liquids gives rise to chemical changes. Michael Faraday called this process **electrolysis**. In the years 1832–4 he established the laws of electrolysis on which the industrial process of electroplating is based.

Fig. 18.2 includes arrows to show the direction of the **conventional current** in a

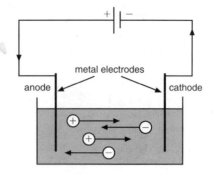

Figure 18.2 *The current is carried by both + ions and − ions in an ionic solution*

circuit. Current is always shown flowing around the circuit **from the cell's positive terminal to its negative terminal**, even though negative ions flow in the opposite direction in the liquid, and negative electrons flow in the opposite direction in the wires. A flow of positive charge in the conventional current direction would, however, have the same effect as the flow of ions and electrons in circuits. Once the conventional current direction is understood, there should be no confusion.

Dilute acid solutions were found to conduct electricity if platinum electrodes were put into the ionised liquid. When the current was passed, water molecules were broken down; hydrogen gas appeared at one electrode and oxygen gas at the other. This process absorbs electrical energy from the power supply which drives the current.

In the **fuel cell** the process has been reversed in order to **produce electrical energy** from the recombination of hydrogen and oxygen gas. This transformation from chemical energy directly into electrical energy with harmless water as a by-product presents technical difficulties, so that fuel cells are not yet widely used.

18.3 Electric charges and currents

The rate of flow of electric charge is called the current. **The current between two points is the charge which flows/time taken**. In symbols this is written as:

$$I = Q/t \tag{18.1}$$

where I is the current in amperes, (A), Q the charge in coulombs, (C), and t is the time taken in seconds, (s). It follows that **amperes are coulombs per second**. A current can be measured by passing it through an ammeter which gives a reading in amps often with a pointer on a numbered scale.

If we want to find the total number of coulombs of charge which flow, we rearrange equation 18.1 so that:

$$Q \text{ (coulomb)} = I \text{ (amp)} \times t \text{ (second)} \tag{18.2}$$

and we can say **charge equals current × time.**

Self test

1 If 15 coulombs of charge flow along a wire in one minute, show that the electric current is 0.25 A, or 250 mA. (Use Equation 18.1).

---

2 What electric currents will transfer:
  (a) 15 C of charge in 30 s?
  (b) 100 C in 50 s?
  (c) 200 C in 1000 s?

  (Answers: 2(a). 0.5 A, 2(b). 2.0 A, 2(c). 0.2 A, or 200 mA).

18.4 Cells and batteries

Chemical cells can be made with two different metal electrodes dipping into various ionic solutions. The chemicals react with one another, causing an electric current to flow through the cells. Thus chemical energy is again transformed into electrical energy as the reactants are changed into chemical products of the reactions. Groups of cells connected together are called **batteries**. They provide the **electromotive force (e.m.f.), which drives electric currents**. We can measure e.m.f. in volts, (V), by connecting a volt-meter to a cell, as in Fig. 18.3. The current driven by a battery always passes in the same

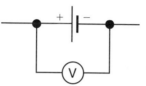

Figure 18.3   A voltmeter can be used to measure the e.m.f. of a cell

direction through the battery. This type of one-way current is called **direct current**.

Both cells and batteries make electric currents flow in networks of conductors. Batteries can be very small where small electric currents are needed. They can fit for example inside a watch, a hearing aid, a calculator, or a Walkman. Larger batteries may:

(a) store more chemicals and,
(b) work for longer before they run down or,
(c) supply larger electric currents.

Rechargeable batteries are those in which the chemical process from reactants to products can be reversed by driving a reverse electric current through the battery. To do this you need a second power supply. A battery charger works on this principle when recharging a car battery which has run down. The charger changes the **alternating current** of the mains electricity supply into direct current suitable for recharging the battery. Of course, electrical energy cannot be created out of nothing, and the mains supplies the energy needed to recharge the battery. In normal practice the car engine has an alternator or dynamo, which recharges the battery while the engine is running

18.5 Electric circuits

To make a steady direct current flow you need **a power supply** like a cell or battery, and you must join its two terminals, (called + and − terminals), to **a continuous closed loop of conductors**. One of these conductors may be a light bulb, a buzzer, or

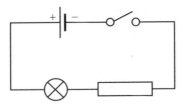

Figure 18.4 The cell provides the e.m.f. which drives the current when the switch is closed

some other device you wish to use. If you also include a switch, this will enable you to turn the electric current on and off when you please. A switch can make or break contact with a loop of conductors, and **no current flows if the loop is not complete**. The complete loop of components is called **a circuit** (see Fig. 18.4).

Insulated copper wires are used to join the circuit components together, because copper conducts electricity very well, and the plastic insulation stops currents being passed to other conductors which may touch the sides of the wires. All the current passes along the copper wires.

18.6 Circuit symbols

It is much easier to describe a circuit using some agreed symbols for the components than to draw what it looks like. Below is a list of symbols which it will be useful to learn.

A cell		Battery	
Normally open switch		Junction of conductors	
Press switch		Fixed resistor	
Normally closed switches		Variable resistor	
Two way switch		Thermistor	
Relay coil		Potentiometer or potential divider	
Relay contacts		Fuses	
Earth		Diode	
Ammeter		Light emitting diode (LED)	

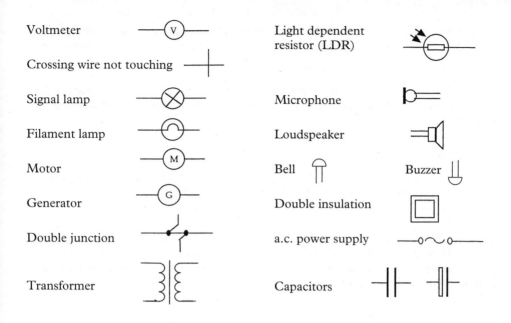

Voltmeter		Light dependent resistor (LDR)	
Crossing wire not touching			
Signal lamp		Microphone	
Filament lamp		Loudspeaker	
Motor		Bell	Buzzer
Generator		Double insulation	
Double junction		a.c. power supply	
Transformer		Capacitors	

18.7 Series and parallel connections

Components connected as in Fig. 18.5a are joined **in series,** which means they are joined in one line so that all the current passing through one component also passes through the other. In series:

$$\text{current } I_1 = \text{current } I_2 \tag{18.3}$$

We connect ammeters in series with components to measure the currents they carry.

Components connected as in Fig. 18.5b are joined **in parallel,** which means part of the supplied current flows through one component, and the rest of the current flows

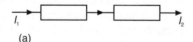

(a)

Figure 18.5   (a) The currents in two components in series are the same, i.e. $I_1 = I_2$

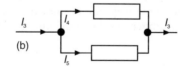

(b)

Figure 18.5   (b) In the parallel connection $I_3 = I_4 + I_5$

(c)   3 A   ?   2.5 A

Figure 18.5   (c) A junction of conductors

through the other one. The two components provide alternative routes for the electricity, and the sum of their two currents is equal to the total supplied current. In parallel:

$$\text{current } I_3 = \text{current } I_4 + \text{current } I_5 \qquad (18.4)$$

It is a general result that when steady currents flow in a network of conductors, **the total current reaching any junction is equal to the total current leaving the junction**.

Self test

Fig. 18.5c shows a junction and currents flowing in two conductors. What is the current in the third conductor? (I = 0.5 A diagonally upwards).

When a cell is connected to a bulb as in Fig. 18.6a, an ammeter on either side of the bulb, at P or Q for example, would read the same. The current does not get less as it delivers energy to the circuit. So in Fig. 18.6a **the rate of flow of electrons in each part of the circuit is the same**.

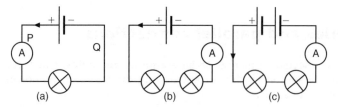

Figure 18.6 *The currents in (a) and (c) are the same, as the ratio of cells to bulbs is the same. In (b) the current is smaller than in (a) or (c)*

In Fig. 18.6b a second identical bulb has been connected in series with the first. The two bulbs shine equally brightly, but they are not as bright as the bulb in Fig. 18.6a. The ammeter shows that the current is less in the circuit with the two bulbs. Adding a third bulb in series would reduce the current still further, and the bulbs would shine even more dimly. It is a general result that more bulbs in series will reduce the current in the circuit, causing the bulbs to shine less brightly.

Fig. 18.6c shows a second identical cell has been connected in series with the first so as to increase the current around the circuit. The two bulbs in the circuit shine as brightly as the single bulb lit by the one cell in Fig. 18.6a. As more cells are connected in series to form a battery, more current flows in the circuit. Provided the cells are all connected to drive current the same way, the current they supply to the series circuit increases with the ratio of the number of cells to the number of bulbs.

18.8 Using voltmeters

Voltmeters as in Fig. 18.3 are connected across the ends of components we study. Connected across the terminals of a cell or battery, a voltmeter measures its e.m.f. in volts. A good voltmeter carries a current so small as to be negligible, and so **a voltmeter connected in parallel with a component** does not noticeably affect the current flowing through it.

We can use a voltmeter to discover that the e.m.f. of a battery is the sum of the e.m.f.s of its separate cells. **With a larger e.m.f., each coulomb of electric charge delivers more energy** to the circuit components. If the e.m.f. of the battery is one volt, then each coulomb of charge delivers one joule of energy as it passes around the circuit. A voltmeter reading tells us the energy change in joules per coulomb as each coulomb passes from one point in a circuit to another. **Volts are joules per coulomb.**

Self test

1 In which circuit of Fig. 18.6 does one coulomb of charge deliver most energy? (Fig. 18.6c)

A difference of V volts between two points in a circuit tells us the change in electrical potential energy of each coulomb passing between the two points. We refer to this difference in potential energy per coulomb as the **potential difference**, (p.d.), between the points. Potential difference, like e.m.f., is measured in volts, or joules per coulomb. To calculate the energy transformed when charge flows between two points, we can use the equation:

Energy transformed (J) = charge (C) × potential difference (V) (18.5)

Example

2 Suppose each cell in Fig. 18.6 has an e.m.f. of 1.20 V. Find the change of energy in joules, when 40 C of charge flow around each circuit.

Answer

2 1.20 V means 1.20 joules per coulomb.
(a) Energy change = 1.20 J/C × 40 C = 48 J.
(b) Energy change = 1.20 J/C × 40 C = 48 J.
(c) Energy change = 2.40 J/C × 40 C = 96 J.
(Note: the two 1.20 V cells in series have an e.m.f. of 2.40 V).

Fig. 18.7 shows a series circuit including a battery and bulbs, to which voltmeters have been added in parallel with the other components.

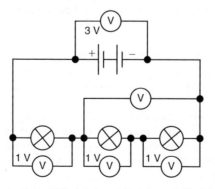

Figure 18.7 A series circuit to which voltmeters have been added. Some voltmeter readings are shown

3 What is the e.m.f. of the battery in Fig.18.7?
4 How much energy is given to each coulomb of charge by the battery?
5 How much of this energy is delivered to each bulb in the circuit?
6 What is the p.d. across each bulb?
7 What is the unlabelled voltmeter reading?
                          (Answers: 3. 3 V; 4. 3 J; 5. 1 J; 6. 1 V; 7. 2 V).

As a general rule, **the e.m.f. of the battery in a series circuit is equal to the sum of the p.d.s across each of the other circuit components**. Each coulomb is supplied by chemical energy in the battery, and delivers this energy to the circuit components. So energy is being transferred from the battery to the circuit; it is not being created or destroyed.

Before leaving this section, it would be useful to note that the insulated copper wires which are excellent electrical conductors do not get very hot. However, the bulb filaments made of tungsten become heated to white heat when they carry the same currents. In general some heat is developed in all resistors carrying electric currents.

Questions

1 (a) How many types of electric charge are there?
   (b) What are they called?
2 What kind of solids make good electrical conductors?
3 Name three materials which may be used as insulators.
4 If a current of 2 A flows in a circuit for 15 s, how much charge in coulombs flows?
5 If 18 C of charge flow from a battery in 6 s, what is the current in amps?
6 Three cells each of e.m.f. 0.8 V are connected in series. What is their total e.m.f.?
7 Three wires, P, Q and S, meet at a junction. A current of 3 A flows in P towards the junction. Q conducts a current of 1.2 A away from the junction.
   (a) How much current flows in wire S?
   (b) In which direction does the current in S flow?
8 (a) How much charge flows when a 5 V battery supplies a current of 0.8 A for 30 s?
   (b) How much energy is delivered to the circuit?
9 A 5 V battery is connected to two components in series. The p.d. across one is 3.5 V. What is the p.d. across the second component?

What have you learnt?

1 Do you know how many types of charge there are?
2 Do you know the difference between electrical insulators like a plastic comb, and conductors?
3 Can you distinguish between electric charge and electric current?
4 Do you know what 'e.m.f.' means, and in what units it is measured?
5 Can you explain what is meant by the p.d. between two points in a circuit?
6 Do you know that in a series circuit the current is the same everywhere along the circuit?
7 Do you know that currents reaching a junction total the same as currents leaving the junction?

8 Can you relate the total p.d. across components in series to the separate p.d.s across each one?

9 Do you know that ammeters are connected in series with components and voltmeters are connected in parallel?

10 Can you draw the symbols for the common electric circuit components?

Using resistors

After reading this chapter you should
- know how resistors and diodes can be used to control current in a circuit
- be able to explain how to measure the resistance of a resistor
- be able to calculate the combined resistance of two resistors in series or parallel
- do calculations involving the e.m.f., current and total resistance in a circuit
- know what determines the resistance of a length of uniform wire
- know how two resistors can act as a potential divider
- be able to do calculations involving power delivered to a circuit
- be able to choose a suitable fuse to protect a circuit.

19.1 Introduction

We have seen that when electric charge passes round a circuit, energy is transformed. For example, electrical potential energy, gained from a chemical reaction in a cell, may be transformed into heat and light in a filament bulb. The energy may be emitted as light in a light emitting diode, (LED), or it may become kinetic energy when the current drives an electric motor. Electric currents are useful in transferring energy from place to place.

Self test

How could you transform electrical energy into sound?

19.2 Controlling the current

One way of controlling the rate of supply of energy to a circuit is to control the size of the current. The bulb shown in the circuit of Fig. 19.1 varies in brightness when a **variable resistor** is adjusted. We saw in chapter 18 that the bulb gets brighter as the current through it increases, so the variable resistor must affect the current. The adjustment is made by sliding a manual control or by rotating a control knob.

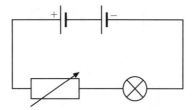

Figure 19.1 *A variable resistor can control the current and the bulb's brightness*

The e.m.f. of a battery also affects the size of the current in the circuit shown in Fig.19.1. In a torch the battery is normally matched with the bulb so that the bulb is at full brightness when connected directly across the battery. A greater current might melt the bulb filament.

Two other circuit components can affect the size of electric currents. One is a light dependent resistor, LDR, which conducts very poorly in the dark, but better when it is well illuminated. The other component is a thermistor, which conducts electricity better when it is gently warmed. Neither the LDR nor the thermistor can carry very large currents as they are both semiconductor devices and not metals. We shall refer to them again in chapter 25.

19.3 Electrical resistance

The property of a circuit which limits the current is its **resistance**. The higher the circuit's resistance, the smaller the current supplied to it by a cell will be. The resistance, R, of the circuit in ohms, (Ω), is the ratio:

resistance in ohms = e.m.f. of cell in volts/current through the cell in amps
In symbols, R (ohm) = V (volt)/I (amp)   (19.1)

It follows that the e.m.f. = current $\times$ circuit resistance, or,
V (volt) = I (amp) $\times$ R (ohm)   (19.2)

> **Example**
> Find the resistance of a circuit in which a cell of e.m.f.
> 3 V drives a current of 0.2 A.
>
> **Answer**
> Use Equation 19.1 where R = V/I = 3 V/0.2 A = 15Ω.

Self test

Show that an e.m.f. of 50 V is needed to supply a circuit of resistance 2 kΩ with a current of 25 mA. (Use Equation 19.2).

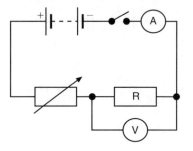

Figure 19.2 A circuit for finding the resistance, R, of a component includes an ammeter in series and a voltmeter in parallel with it

19.4 Individual resistors

Fig. 19.2 shows a circuit which can be used to measure the resistance, R, of a single conductor. Equation 19.1 defines the resistance of any conductor, where V is the p.d. across it, and I is the current through it. Resistances can vary from fractions of an ohm to millions of ohms, (megohms, $M\Omega$). Notice that the voltmeter in Fig. 19.2 is connected in parallel with the resistor being studied. The purpose of the variable resistor is to vary the current in the whole circuit, so that the resistance R may be found for a range of currents. Table 19.1 gives a series of possible readings, and Fig. 19.3 shows the results plotted on a graph.

Table 19.1

I(A)	V (V)	R = V/I(Ω)
0	0	—
0.1	2.2	22
0.3	6.6	22
0.5	11.0	22
0.7	15.4	22

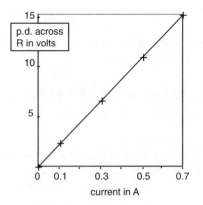

Figure 19.3 Table 19.1 has typical data obtained with a fixed resistor, R, using the circuit of Fig. 19.2. Here, we see the p.d./current characteristic. The straight line graph has a slope of 22 V/A, so the resistance, R, is 22Ω.

Table 19.1 shows that the conductor has a fixed resistance of 22Ω. This is confirmed by the linear graph whose gradient, V/I, is constant and equal to the resistance. If the resistance had been greater, the graph would have been steeper.

Conductors whose resistance remains constant for different currents are called **ohmic conductors**, because they obey **Ohm's Law**. **This law says that the ratio (p.d. across a conductor/current through it) is constant, provided the conditions like temperature remain constant**. Metals and some other conductors are ohmic. For some conductors, the ratio (p.d./current) is not constant, and these are said to be non-ohmic.

When the filament of a bulb becomes white hot, its resistance increases. This is illustrated by the graph of p.d. across the filament plotted against the current through it (Fig. 19.4). As the filament becomes hotter, the graph curves towards the V axis, and

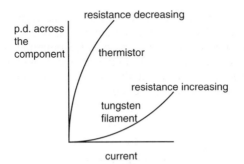

Figure 19.4 As currents increase in a thermistor and a bulb filament, their temperatures rise. The resistance of the semiconductor falls, but the metal's resistance rises. So both graphs curve

the ratio V/I for the metal increases. The vibrations of molecules in the metal obstruct the flow of electrons more at high temperatures than at lower ones, and this causes the increase in the metal's resistance.

It is possible to distinguish a semiconductor from a metal because the metal's resistance increases as it becomes hotter, whereas the semiconductor's resistance decreases as it warms up (see Fig. 19.4).

19.5 Diodes

Diodes are made of two different semiconductor materials joined in series. The flow of charge across the boundary between the two materials is easy in one direction, but very difficult in the reverse direction.

Diodes have very high resistance to electric current in one direction, and low resistance to current in the opposite direction. The voltage/current characteristic for a typical diode in Fig. 19.5 shows that it is not an ohmic conductor, so the ratio (of p.d. across it/current through it) is not a constant. A protective resistor is normally used to keep down the current through a light emitting diode. Many LEDs work with currents of a few milliamps.

In circuits with diodes, it matters which way round the diodes are connected. The **circuit symbol for a diode includes an arrow showing in which direction a conventional current will flow through the device**. The thick straight line at the tip of

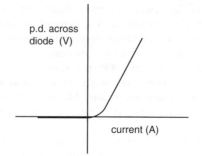

p.d. across
diode (V)

current (A)

Figure 19.5 *The voltage/current characteristic for a diode*

the arrow in the circuit symbol may be thought of as representing a barrier to current in the opposite direction. The conventional current direction in a conductor is from the more positive potential towards the less positive potential. We may think of currents emerging from the red positive terminal of a cell, and returning to the black negative terminal after flowing through the circuit components (see Fig. 19.6). We shall discover some uses for diodes in Section 25.9

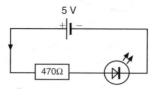

5 V

470Ω

Figure 19.6 *A protective resistor of 470Ω limits the current. The LED is connected the right way round*

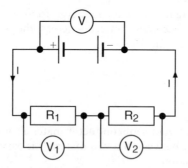

V

R_1

V_1

R_2

V_2

Figure 19.7 *Voltages in series 'add up'*

19.6 Resistors in series

In Fig. 19.7 the p.d.s across the two resistors add up to the e.m.f. of the cell (see chapter 18). In symbols $V = V_1 + V_2$, and so we may write:- $I \times R = I \times R_1 + I \times R_2$, where R is the total resistance of the circuit. Dividing this equation by the common factor I, we obtain:

$$R = R_1 + R_2 \qquad (19.3)$$

This tells us that **the total resistance of conductors in series is the sum of their separate resistances**.

We can apply this result to the resistance of a length of uniform wire. Since each metre of the wire has the same resistance, the total resistance is the sum of the resistances of each metre of the wire. So the **total resistance is proportional to the total length of the wire**.

Self test

Find the current in a circuit consisting of a cell of e.m.f. 2.5 V connected to a 15Ω resistor in series with a bulb of resistance 10Ω.   (I = 0.1 A)

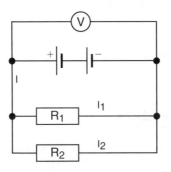

Figure 19.8

19.7 Resistors in parallel

In Fig. 19.8 the current I is equal to the sum of the currents in the parallel resistors. In symbols: $I = I_1 + I_2$, and so we may write: $V/R = V/R_1 + V/R_2$, where R is the circuit's resistance. Dividing this equation by the common factor V, we obtain:-

$$1/R = 1/R_1 + 1/R_2 \qquad (19.4)$$

This can be rewritten for **two conductors in parallel** as:-

$$R = (R_1 \times R_2)/(R_1 + R_2) \qquad (19.5)$$

Self test

Show that two 30Ω resistors in parallel act like one resistor of 15Ω.

19.8 Uniform wire resistors

A thick wire resistor conducts better than a thinner one of the same length. Two identical resistors connected in parallel combine to produce half the resistance of a single resistor. In the same way a uniform wire with double the area of cross section has half the resistance of the original wire. The resistance of a wire can be written:

$$\mathbf{R} = \mathbf{s} \times \mathbf{l}/\mathbf{A} \tag{19.6}$$

where the wire has length l metres, area of cross section A square metres, and is made of material whose **specific resistance, (resistivity),** is s ohm.metres.

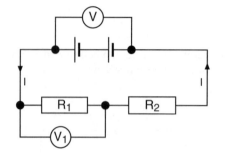

Figure 19.9 Two resistors act as a potential divider

19.9 Potential dividers

In Fig. 19.9 the potential difference V_1 across the resistor R_1 is a fraction of the e.m.f. V of the cell. So resistors can be chosen to step down an applied e.m.f. to make a smaller p.d. In the circuit:

$V_1 = I \times R_1$, and
$V = I(R_1 + R_2)$.

If we divide the first equation by the second we find the ratio: $V_1/V = R_1/(R_1 + R_2)$. This leads to the result:

$$V_1 = V \times R_1/(R_1 + R_2) \tag{19.7}$$

Potential divider circuits, or **potentiometers**, are used widely in electronic controls. For example the arrangement can be used as a volume control for a sound system (see Fig. 19.10). As the sliding contact on the right is moved upwards, the p.d. of the signal to be amplified increases. The power output to a loudspeaker increases, and the sounds become louder.

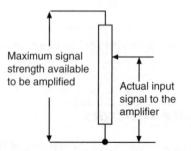

Maximum signal strength available to be amplified

Actual input signal to the amplifier

Figure 19.10 The potential divider controls the p.d. of a signal to be amplified in a sound system

19.10 Electrical power transfer

Power is the rate of transforming energy. As we saw in chapter 6, power is measured in joules per second, J/s, or watts, W.

Power = energy transformed/time taken. In symbols:
P (watt) = V (volt) × Q (coulomb)/t (second)   (19.8)

since the energy transformed = p.d. × charge (see chapter 18). But since Q/t is the current in amps:

P (watt) = V (volt) × I (amp)   (19.9)

and so the **power delivered is the p.d. × current**.

19.11 Heating power in a resistor

Since the p.d. across a resistor, V = I × R, equation 19.9 can be put in the form:-

$P = I^2 \times R$ (19.10)

It follows that the power transformed in a resistor is proportional to the (current)^2. So if the current is doubled, the power transformed will be quadrupled, etc.

___ **Self test** _____

Fill in the gaps to complete the table below.

	p.d. across resistor	current	resistance	power
	V (V)	I (A)	R (ohm)	P (watt)
a.	10	0.2		
b.	240		120	
c.		4		1000

Answers:
(a) R = 50Ω, P = 2 W. (b) I = 2 A, P = 480 W, (c) V = 250 V, R = 62.5Ω.

Using Equation 19.10 we can see that for a certain current, the power transformed in a resistor is in proportion to its resistance, R. **If R is negligible, so is the power P**; so copper wires which have low resistance do not get hot in normal working conditions. They have much less resistance than a bulb filament, which does get very hot when it conducts the same current.

___ **Self test** _____

Find (a) the current, and (b) the resistance of a 60 W filament bulb, joined to a 240 V mains supply.   (Answers: (a) I = 0.25 A, (b) 960Ω).

19.12 Protecting electrical circuits

A faulty connection may lead to a short circuit resulting in a very big current. Large currents can seriously damage the wiring or components of a circuit, and to prevent this some circuits include a series resistor called **a fuse**. The fuse melts when the current exceeds a certain safe value. When the fuse melts, the current is turned off and damage is prevented. The fuse can be replaced at little cost when the faulty circuit has been repaired.

An alternative to the fuse is a magnetic switch which cuts out when the current becomes too large. The switch is closed manually to reconnect the circuit, and costs nothing to operate. Magnetic relay switches replace fuse boxes in some domestic situations, where they provide the same protection for the wiring as fuses, and are more easily turned on again after repair.

Choosing the correct fuse requires that you know the current normally taken by each appliance. Mains equipment running off 240 V are labelled to show power consumption in watts. Equation 19.9 can be used to calculate the operating current:

$$I \text{ (A)} = \text{power } P \text{ (W)/voltage } V \text{ (V)} \tag{19.11}$$

Then the fuse must be chosen so that it will melt only if a current significantly higher than the expected value flows in the circuit. So, for example, lighting circuits may have a 5 A fuse even though the current is unlikely to be higher than 2 A. Power circuits have thicker wire with thicker insulation. They may normally carry 10 A, so 13 A fuses are used rather than 5 A fuses which would be useless.

Self test

(a) Show that a 2 kW washing machine takes a current of more than 8 A from a 240 V supply.
(b) Choose a fuse for the mains plug of the machine, if you have 1 A, 3 A, 5 A, 8 A, and 13 A fuses to choose from. (Use Equation 19.9a. I = P/V = 8.3 A, b. 13 A fuse).

Questions

1 Resistors of 30Ω and 20Ω are joined in series with a battery of e.m.f. 5 V. Find:
   (a) the total resistance of the circuit,
   (b) the current in the battery, and
   (c) the p.d. across each resistor.
2 A potential divider or potentiometer has a 9 V battery connected across its ends. A voltmeter is connected across one third of the potentiometer's resistance.
   (a) Draw this circuit arrangement.
   (b) Write down the voltmeter reading.
3 20Ω and 30Ω resistors are joined in parallel and a cell of e.m.f. 2 V is connected across them.
   (a) Draw the diagram for this circuit.
   (b) Find the current in each resistor.
   (c) Calculate the circuit's resistance.
4 Fig. 19.11 shows three voltage/current characteristics.
   (a) Which graph relates to an ohmic conductor?
   (b) Which graph refers to a semiconductor?
   (c) Which graph refers to a tungsten bulb filament?

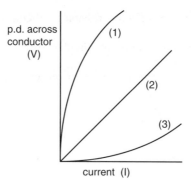

Figure 19.11   P.d./current for three components of electric circuits

5   A single bar electric fire of power 1 kW works with a mains supply of 240 V.
   (a) Make a rough estimate of the current flow.
   (b) Estimate a suitable rating for the fuse in its plug.
   (c) Estimate the resistance of the heating element.
   (d) Given a uniform wire of resistance 40Ω/m, estimate the length of this
        wire needed for the element of the fire.
6   (a) What happens to a fuse in a circuit as the current become very large?
   (b) Why are fuses used in electric circuits?
7   An ohmic conductor carries a current of 2 A and transforms 400 J/s into heat
    energy.
   (a) What is the p.d. of the supply in volts?
   (b) What is the resistance of the conductor?
   (c) What p.d. will be needed to double the current to 4 A in this conductor?
   (d) What will be the power transformed when the current reaches 4 A?
8   A length of uniform wire has a resistance of 5Ω. What will be the resistance of
    a wire of the same material if:
   (a) it has the same cross-section but three times the length of the original
        wire?
   (b) it has the same length but one quarter the area of cross-section of the ori-
        ginal wire?
   (c) it has three times the length and one quarter the area of cross-section of
        the original wire?
9   Name a semiconductor device which conducts electricity better in the light
    than in the dark.

What have you learnt?

1   Can you draw a circuit diagram to show how you would find the resistance
    of a resistor?
2   Can you explain how to calculate a resistance from experimental data?
3   Do you understand that resistors transform electrical energy into heat?
4   Can you distinguish the different voltage/current characteristics for different
    conductors?
5   Can you write down the resistance and power transformed in part of a cir-
    cuit, given the p.d. and the current?
6   Can you write down the resistance of a circuit having a number of resistors
    joined in series?

7   Can you write down the effective resistance of conductors arranged in parallel?
8   Do you understand how a potential divider works?
9   Do you know what a diode is?
10 Do you know how to choose a fuse to protect a piece of electrical equipment?

20 Magnetic forces

Objectives

After completing this chapter you should
- know how a magnetic compass works
- understand how to describe a magnetic field, by showing its lines of force
- understand that magnets exert forces on each other
- be able to describe a solenoid and an electromagnet
- know the difference between hard and soft magnetic materials
- be able to explain how an electric bell works
- know that a wire carrying current across a magnetic field experiences a force.

20.1 Introduction

If you have played with magnets you will probably be familiar with two of their basic properties. First, that they attract pieces of iron or steel, and secondly, that if they are suspended freely they always come to rest pointing in a definite direction. The end of a suspended magnet which points towards the North of the Earth is called the **North-seeking pole** or simply the **N-pole**. The end which points South is called the **South-seeking pole** or **S-pole.**

Each magnet has both a North-seeking and a South-seeking pole. If you try to separate them by cutting a magnet in half, you end up with two complete magnets each with a North-seeking and a South-seeking pole. The magnetism appears concentrated at the two poles. Fig. 20.1 shows how small tacks are attracted to both the magnetic poles when a bar magnet is dipped into a box of iron tacks. Two other substances that have the same magnetic properties are cobalt and nickel.

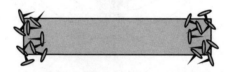

Figure 20.1   A bar magnet attracts small iron tacks to both its ends, where magnetism appears to be concentrated

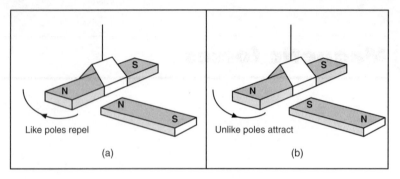

Figure 20.2 *Magnetic forces acting on suspended magnets cause their deflections*

20.2 Forces between magnets

Tests with two magnets show that when two **like poles** are brought close together, they **repel each other**; but when two **unlike poles** are brought close, they **attract each other** (see Fig. 20.2). The forces increase when the magnets are brought closer and decrease as they are separated. We can tell that the Earth must have a South-seeking magnetic pole in the Arctic regions, as it attracts the unlike North-seeking poles of compasses and repels their South-seeking poles.

20.3 Magnetic fields

The region around a magnet where we can detect its influence is known as a **magnetic field**. One way of charting the magnetic field near a magnet is to use fine iron filings scattered on paper above the magnet (see Fig. 20.3). When the paper is gently tapped, the filings settle in a symmetrical pattern of lines linking the two magnetic poles.

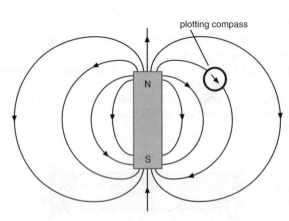

Figure 20.3 *The magnetic field pattern around a bar magnet can be seen with iron filings, or with a small plotting compass placed in the field*

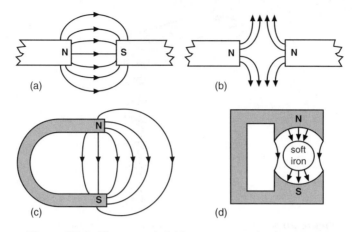

Figure 20.4 *The magnetic field patterns around various magnets*

A second method uses a small plotting compass which points along a **magnetic line of force** where it is placed on the paper. The pattern of these lines is usually represented on paper with an arrow on each line to show the direction in which the North-seeking end of a compass would point.

Fig. 20.4 shows the patterns of magnetic lines of force in the air gaps around variously shaped magnets. Magnetic lines of force do not cross. If they did, it would indicate two possible compass directions at a single point, and that is never the case. Where the lines of force are closer, the field is stronger; and where they spread out, the field weakens. The lines seem to behave like invisible elastic bands pulling inwards on unlike magnetic poles.

20.4 Induced magnetism

When an iron nail is set down near a permanent magnet, it too may become a magnet with a North-seeking and a South-seeking magnetic pole. A second nail set end to end with the first one may also become magnetised in the same way. **The magnetism is said to have been induced in the nails** as they lay along magnetic lines of force.

An iron nail held near one end of a bar magnet is always attracted towards it. This shows that the induced magnetic pole at one end of the nail is always of opposite kind to the nearby inducing pole in the magnet. Fig. 20.5 illustrates induced magnetism in several iron tacks. Near the permanent magnet their induced magnetism enables each tack in the chain to be supported by the one above. Each tack has a magnetic pole at the top unlike the pole induced at the bottom of the supporting tack. When the permanent magnet is removed, the tacks drop off the chain, having lost most of their induced magnetism.

20.5 Magnetic materials

Groups of iron atoms form tiny magnets. When their polar axes are not aligned in any particular direction, the material containing them is not magnetised. To magnetise the iron we need to treat the atomic groups like compass needles which all point in a

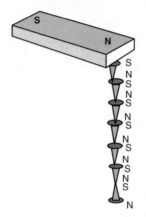

Figure 20.5 Induced magnetism in tacks enables a chain to form

certain direction in a strong magnetic field (see Fig. 20.6). As the groups' polar axes are lined up, the material becomes magnetised; and when the magnetising field is removed, some of the atoms may remain lined up.

Materials which easily gain and lose their magnetism are called **'soft' magnetic materials**. A typical example is the soft iron from which iron tacks and nails are made. Materials that are difficult to magnetise or demagnetise are called **'hard' magnetic**

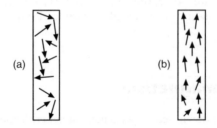

Figure 20.6 (a) An unmagnetised iron bar has small magnetised zones with magnetism directed in random directions. In (b) these directions have been aligned to create a magnet

materials. A permanent magnet must be made of a hard magnetic material so as to remain a magnet. Some steels are magnetically hard, and an alloy of aluminium, cobalt, and nickel (known as 'AlNiCo'), is used for some bar magnets.

Horseshoe-shaped magnets have their poles close together. This results in a stronger magnetic field in the air between the poles. When not in use the magnet's poles are joined with a piece of soft iron called a **keeper** (see Fig. 20.7). The magnetism induced in the keeper helps the magnet to retain its magnetism.

20.6 Test for magnetism

To test whether a piece of metal is a magnet, one tests for **repulsion** between the metal and each pole of a known magnet. If there is attraction between the test piece and a magnet, we can deduce that the test piece contains a magnetic material, but it

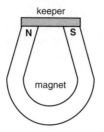

Figure 20.7 A horseshoe magnet is stored with its keeper between its poles

may not be a magnet; its magnetism may have been induced. For repulsion to occur between the test piece and a magnet however, they must both be magnets.

20.7 Magnetising a piece of iron

(a) Molecular vibration. We could line up an iron bar in a strong magnetic field and hammer the bar to shake up its molecules. The field would tend to line up the polar axes of the atomic groups so that more of them would lie along the field's direction.

(b) Using the magnetic effect of an electric current. In Fig. 20.8a direct current circuit is wrapped around a six inch nail of soft iron to create **an electromagnet.**

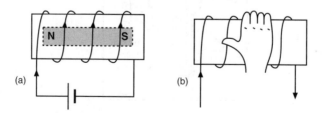

Figure 20.8 An iron nail is magnetised by a current in a coil of insulated wire. The nail induces magnetism in drawing pins, but when the current is turned off, the pins fall from the nail

When the current is turned on, the nail becomes a magnet; and when the current is turned off, the nail loses nearly all its magnetism.

Fig. 20.9a shows an iron bar being magnetised inside a hollow tube of cardboard around which a coil of insulated copper wire has been wound many times. The current

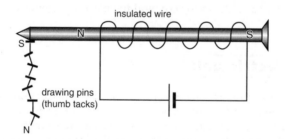

Figure 20.9 The right hand's thumb is used to show where the N-seeking pole occurs inside a solenoid. The fingers follow the current direction

in the coil or **solenoid** generates a magnetic field like the field of a bar magnet. The field also continues inside the coil along its axis, so magnetic lines of force are shown to be continuous closed loops. This is another example of an electromagnet.

We can increase the strength of the magnetic field inside a solenoid by:

(a) increasing the current
(b) winding more turns on top of the original ones
(c) putting soft iron inside the solenoid.

Fig. 20.9b shows how to remember which end of the magnetised bar behaves as a North-seeking pole. If you grip the solenoid in your right hand with fingers following the current direction, your extended thumb shows the end where there is a North-seeking pole.

An electromagnet with a soft iron core can be used in a scrap yard to separate metals containing iron from non-magnetic materials like aluminium, glass, wood, plastic and copper.

20.8 Demagnetising an iron bar

Alternating current in a solenoid switches direction many times each second. When its direction reverses, so does its magnetising effect. So the magnetism of an iron bar inside the coil reverses its direction when the current reverses. If the bar is now gradually removed from the coil while the alternating current continues to flow, the magnetising effect on the bar reduces to zero, and the bar is demagnetised.

20.9 The electric bell

Fig. 20.10 shows five different materials used in the construction of a bell. When a push switch is pressed, an electromagnet with coils of insulated copper wire (1), wound on a soft iron core (2), attracts a soft iron armature so that a small steel ball hits a steel gong (3). A springy brass strip (4) allows the armature to move, but contact is

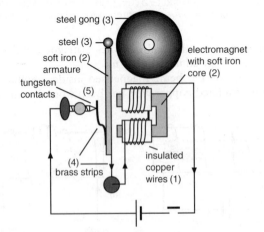

Figure 20.10 A 'make and break' contact at (5) enables the bell to keep ringing while the switch is pressed

broken at (5), where a screw with a tungsten tip can contact a second strip of brass. Sparks sometimes occur at the contact points. The tungsten withstands high temperatures from the sparks and hence wears well. When the contact is broken, the current stops, the soft iron is no longer attracted to the electromagnet, and the first brass strip returns the armature to its starting place. This turns on the current and the cycle is repeated. Steel is used for the gong as it is hard and makes a good ringing sound.

20.10 Magnetic effects of electric currents

A current in a single straight electrical conductor generates a magnetic field in the space nearby. Fig. 20.11 shows plotting compasses on a card in a plane at 90° to the conducting wire. The compasses show that the magnetic lines of force are a series of

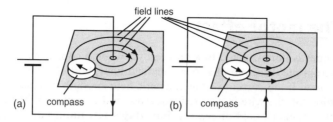

Figure 20.11 (a) Magnetic lines of force encircle the wire carrying an electric current. (b) The compass directions reverse when the current is reversed. A right-handed corkscrew, advancing with the current, rotates as indicated by the magnetic field arrows

circular rings centred on the conductor. The directions of the compasses reverse when the terminals of the d.c. supply are swapped over and the current flows in the opposite direction. If you imagined gripping the conductor in your right hand with outstretched thumb following the current direction, your fingers would indicate the direction of the field lines around the conductor.

Magnetic fields generated near permanent magnets can be produced just as effectively by means of currents in insulated wires. The strength of the field around any conductor carrying an electric current can be increased by filling the space with soft iron. This explains why soft iron is used so much as the core in transformers, electromagnets and electrical appliances like bells, motors and dynamos.

20.11 Forces when magnetic fields are distorted

When one magnet is brought close to another, both their magnetic field patterns are distorted and a force exists between the magnets.

Suppose two electrical conductors, each carrying a current, were brought close to one another so that both their magnetic field patterns were distorted. What would happen then?

Fig. 20.12 shows the general result for two parallel conductors. **If their currents are in the same direction, the conductors attract each other, but when the currents are opposed, the conductors repel each other**. This also applies to two circular loops of wire on a common axis. The force on each loop increases with each of

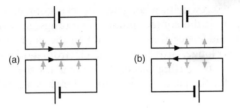

Figure 20.12 *The grey arrows indicate forces on conducting wires. In (a) parallel currents attract each other; in (b) antiparallel currents repel*

the currents. On a long solenoid where the currents flow continuously in the same sense, each turn of wire attracts its neighbours.

20.12 The motor effect

A piece of copper is not attracted to a magnet, yet when a copper wire carries an electric current across the direction of a magnet's field lines, both the wire and the magnet experience a force. If the current or the magnetic field strength increases, the force gets bigger. The force on the wire is directed sideways, and acts at right angles to both the current and the magnet's field directions. **Faraday's left hand rule** states that, if the first finger of the left hand indicates the direction of the magnetic field, and the second finger is parallel to the current, then the thumb indicates the direction of the force on the conductor causing its motion (see Fig. 20.13). This is known as **the motor effect**.

The motor effect force is strongest when the current and the field directions are at 90°, but there is no force when the two directions are parallel. The force on the conductor is reversed when the direction of either the field or the current is reversed. Applications of the motor effect force include electric motors, loudspeakers and moving coil galvanometers. The next chapter explains how they work.

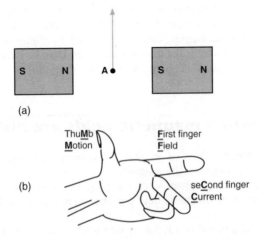

Figure 20.13 *(a) When a conductor at A carries current towards the reader, it is forced upwards in the magnetic field of the magnets. In (b), the left hand indicates the directions of the magnetic field, the current and the conductor's motion*

1 Draw a diagram showing the magnetic lines of force around a bar magnet.
2 Two bar magnets are side by side. They attract each other. Draw a diagram of the magnets and label their poles.
3 Some permanent magnets are made of an alloy called AlNiCo containing atoms of aluminium, nickel and cobalt. Which of these elements are magnetic materials?
4 Why is soft iron suitable for the core of an electromagnet?
5 A solenoid carries direct current and a six inch nail is held along its axis, first just beyond one end, and then just beyond the other end. Draw diagrams showing the direction of the force on the nail in each situation.
6 Two identical solenoids are placed end to end. How could they be connected in series so that they repel each other when a battery drives direct current through both coils? (Draw a diagram.)
7 A copper wire carries a current across a perpendicular magnetic field.
   (a) How does this affect the wire?
   (b) What happens if (i) the current is reversed, (ii) both the current and field are reversed?
8 A current in a coil sets up a magnetic field. What three changes would increase the strength of the field?
9 Draw a diagram of an electric bell and label its parts.

What have you learnt?

1 Do you know which three chemical elements are strongly magnetic materials?
2 Can you explain induced magnetism?
3 Do you know what affects the direction indicated by a compass needle?
4 Can you describe how to make an electromagnet?
5 Do you know how to demagnetise an iron rod?
6 Can you describe the forces between magnets?
7 Do you know how to increase the magnetic field inside a solenoid?
8 Can you describe the motor effect force on a conductor carrying a current in a magnetic field?
9 Can you explain the difference between hard and soft magnetic materials?
10 Can you explain how an electric bell works?

21) Applying the motor effect

21.1 Introduction

Section 20.12 described the motor effect force on a single conductor in a magnetic field. In Fig. 21.1a we are looking down on a rectangular loop of wire between the poles of a magnet. The arrows show that opposite sides of the coil carry current in opposite directions, so that the left side is pulled towards the reader while the right side is forced into the page. Fig. 21.1b shows the same circuit, as seen from the cell at the end. On the right the wire carrying current away from you is forced down, but on the left the wire carrying current towards you is forced up.

The loop of wire in Fig. 21.1 could be mounted on a central axis parallel to its

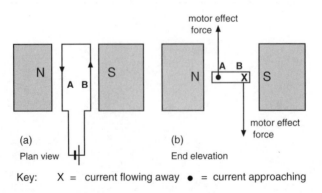

Key: X = current flowing away   • = current approaching

Figure 21.1   (a) A plan view of a circuit between magnetic poles. (b) shows the end view seen from the cell, including the motor effect forces on opposite sides of the coil

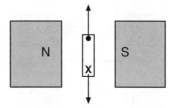

Figure 21.2 *When the circuit in Fig. 21.1b swings into a vertical plane, the forces cause no further rotation*

longer sides. Then it would rotate until one side of the coil was at its highest point, and the other side at its lowest point (see Fig. 21.2). The coil would then quickly stop rotating. If we reversed its current, the side of the loop at the top would be forced down and the side at the bottom would be forced up. By reversing the current we could cause the loop of wire to rotate a further half turn. Then the coil would stop again. In order to keep the loop rotating we would have to reverse the direction of its current after every half turn. However there is another problem which needs to be overcome. That is, the wires between the cell and the loop get twisted up when the loop rotates. This must be avoided in a motor.

21.2 The main parts of a d.c. electric motor

The part which rotates in an electric motor is called the **armature**. It is made of laminated soft iron and it is mounted on bearings at each end which allow it to rotate freely. The armature fills much of the gap between the poles of a **permanent magnet** and so a strong magnetic field exists in the gap. The faces of the magnet's poles are shaped to make this field radial and uniform in strength as was indicated in Fig. 20.4d.

An **armature coil** of insulated wire is wound onto the soft iron. A bigger turning effect can be produced with more turns on this coil, as each turn experiences motor effect forces.

At each end of the armature coil, a short length of insulation is removed from the wire and a connection made to one of two curved brass or copper strips, insulated from each other and from the rest of armature. Fig. 21.3 shows the **brushes** which

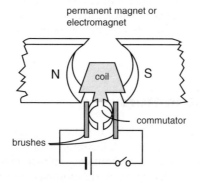

Figure 21.3 *The brushes and commutator supply current to the armature coil, and reverse the current in the coil every half turn*

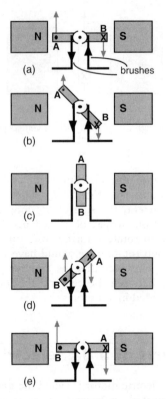

Figure 21.4 A split ring commutator connects the brushes to the armature coil of a motor. (a) to (e) above show how forces on the coil maintain its rotation

conduct current into and out of the armature coil via the curved strips. The two metal strips form a **commutator**, which is a part of the armature. There are two gaps between the semicircular commutator strips shown in Fig. 21.3. As the armature rotates, the gaps pass the brushes every half turn, and this reverses the direction of current in the coil without changing its direction from the power supply (see Fig. 21.4). The commutator's function avoids the twisting of the supply wires; it also allows the brushes to be positioned so that the current in the armature coil is reversed when it needs to be. This keeps the armature rotating continuously.

The armature current is biggest when the power supply is first turned on, and the faster its coil rotates, the smaller its current becomes. We shall say more of this in Section 24.3.

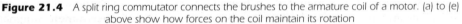

21.3 a.c. electric motors

We have described electric motors which use d.c. supplies, but **a.c. electric motors are also widely used**. Fig. 21.5 shows a locomotive power car found at the front and rear of each Eurostar train (see also the front cover illustration).

The locomotive is driven by a.c. electric motors, one motor to each wheel axle. The nearest wheel axles in the adjoining cars are also similarly powered, so that a total of 12 axles on the train are driven like those shown in Fig. 21.5.

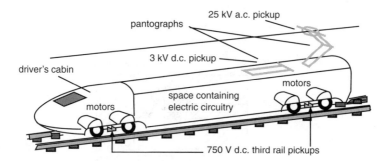

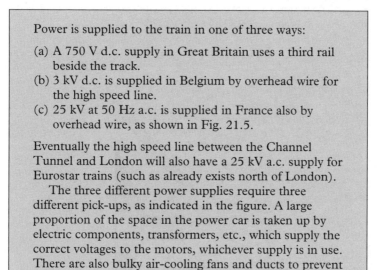

Figure 21.5 The front and rear cars of a Eurostar train each have four driven axles. The four a.c. electric motors take up little space. Note the three types of pick up for the different types of electricity supply

> Power is supplied to the train in one of three ways:
>
> (a) A 750 V d.c. supply in Great Britain uses a third rail beside the track.
> (b) 3 kV d.c. is supplied in Belgium by overhead wire for the high speed line.
> (c) 25 kV at 50 Hz a.c. is supplied in France also by overhead wire, as shown in Fig. 21.5.
>
> Eventually the high speed line between the Channel Tunnel and London will also have a 25 kV a.c. supply for Eurostar trains (such as already exists north of London).
> The three different power supplies require three different pick-ups, as indicated in the figure. A large proportion of the space in the power car is taken up by electric components, transformers, etc., which supply the correct voltages to the motors, whichever supply is in use. There are also bulky air-cooling fans and ducts to prevent the circuits from overheating.

21.4 The moving coil galvanometer

A moving coil galvanometer is a sensitive current detector. When a current flows in its coil, the coil begins to rotate between a magnet's poles, as if it were part of an electric motor. But the galvanometer has no commutator. The coil's rotation starts to twist the spiral springs to which the coil is attached, one of which is shown in Fig. 21.6. The rotation ceases when the couple exerted by electromagnetic forces on the coil is just balanced by the restoring couple exerted by the twisted spiral springs. The moving coil galvanometer has a pointer attached to its coil, and there is a numbered scale to show how far the pointer has moved, and the coil rotated.

The springs return the pointer to the zero position on the scale when no current is flowing. They exert stronger restoring forces on the coil the further it rotates. In this way it takes a larger current to move the pointer further around the scale. So the pointer's deflection on the scale indicates the size of the current in the coil.

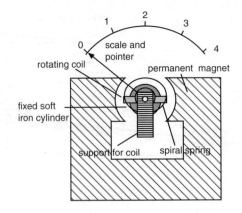

Figure 21.6 A moving coil galvanometer has no commutator. Electrical connections are made between the circuit and the coil via springs. Each spring is attached to a fixed support at one end, and to the coil at the other end.

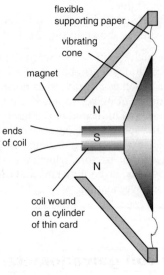

Figure 21.7 (a) A loudspeaker

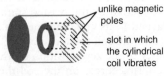

Figure 21.7 (b) The permanent magnet has a slot for the loudspeaker's coil

21.5 Loudspeakers

In a loudspeaker a coil with turns of thin copper wire is wound on a small cylinder of cardboard. The coil carries alternating current. Inside the coil is a pole of a permanent magnet; the other pole is outside the coil across a narrow air gap. The thin cardboard tube which carries the coil can move freely between the magnetic poles (see Fig. 21.7).

When the current flows in one direction in the loudspeaker's coil, there is a motor effect force on the coil parallel to its axis. When the current direction is reversed, the motor effect force is also reversed; so the coil vibrates along its axis when it carries a.c. The vibrations have the same frequency as the a.c. in the coil.

One end of the cardboard tube is joined to the apex of a cardboard cone, so that the cone vibrates with the cylinder. The open, wide end of this cone is supported by flexible corrugated paper attached to a metal frame. When the cone vibrates, air in front of it is alternately compressed and decompressed so that a sound is emitted.

The sound has the same frequency as the current in the loudspeaker's coil. Its loudness increases with the size of the alternating current, because a greater current causes the cone to move with larger amplitude vibrations.

Questions

1 Explain with a diagram how a commutator and brushes are used in an electric motor.
2 Why does the armature of a d.c. motor rotate?
3 What is the purpose of springs attached to the coil of a moving coil galvanometer?
4 How would you use a galvanometer to find which of two currents is the larger?
5 If the springs in a galvanometer were made stiffer how would this affect its readings?
6 (a) What type of current is needed in a loudspeaker to produce sound?
   (b) What change is needed to make the loudspeaker sound a note of higher pitch?
   (c) What change is needed to make the loudspeaker produce a louder sound?
7 How could the forces which rotate the armature of a motor be increased?
8 What sort of material is best for the armature of a motor?

What have you learnt?

1 Can you describe the motor effect force on a conductor carrying a current in a magnetic field?
2 Can you name the parts of a d.c. motor and explain their functions?
3 Do you understand how to increase the turning effect of a motor's armature?
4 Do you understand the functions of parts of a moving coil galvanometer?
5 Do you understand how a loudspeaker works?
6 Do you understand how to vary the loudness and pitch of sounds emitted by a loudspeaker?

22 **Electricity in the home**

22.1 a.c. and d.c.

The mains supply to our homes is an alternating current (a.c.) supply. Fig. 22.1 shows a graph of two cycles of the p.d. across the supply plotted against time. In Britain each cycle of the mains takes (1/50) s, so that there are 50 complete cycles per second. It is a 50 Hz supply, generated in power stations where the frequency is determined by the rate of rotation of magnets driven by steam turbines (see Section 24.4).

The p.d. across an a.c. supply is continually changing; so why is it labelled

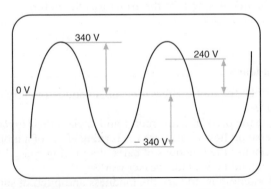

Figure 22.1   A graph of two cycles of an a.c. 240 V supply. The peak voltage is 340 V, but the average power delivered is equivalent to having a steady 240 V d.c. supply

'240 V'? The answer is that **an a.c. supply of 240 V delivers energy to a resistor at the same average rate as would a steady 240 V d.c. supply**. So by labelling an a.c. supply in this way we can compare a.c. and d.c. supplies. The same system is used when we talk of the size of an alternating current in amps. Two amps a.c. delivers energy to a resistor as fast as two amps d.c. While much of Europe has a 240 V, 50 Hz a.c. supply, in North America the supply is 110 V and its frequency is 60 Hz.

In order to deliver energy at the same average rate as a 240 V d.c. supply, an **a.c. 240 V supply** must have a **peak voltage** of more than 240 V (see Fig. 22.1). The insulation of the wires which carry the mains supply must be able to stand up to p.d.s of more than 340 V, so that the supply wires are effectively insulated, even at the peak value of the p.d. in each cycle.

22.2 Safety and domestic supplies

Before mending a tap washer, you must turn off the water supply at the place where it enters your home. Similarly **before examining your electric wiring, you must turn off the electricity supply** at the fuse box where the supply wires enter your home. The wires could give you an electric shock. So plugs, sockets, and cables are all enclosed in tough insulating materials which are perfectly safe to touch in normal use.

Your muscles are normally controlled by small currents carried by the nerves from your brain. Nerves can be damaged by larger currents, and the risk arising from contact with a mains wire is greater when your skin is wet. This is why light switches in bathrooms are often set in the ceiling, where they cannot be touched. They are operated remotely by cords of insulating material.

The fuse in a lighting circuit is not there primarily to protect people from electric shocks. It is there to protect the wiring in the walls from becoming overheated and presenting a fire risk.

22.3 Why have an earth wire?

The mains electricity is carried in two wires called **live and neutral**. At first sight a third wire, called the **earth** wire, appears to serve no purpose, since it is supposed to carry no current. But it provides an **essential safety feature for people using an appliance with a metal case**, such as a washing machine, tumble drier, refrigerator, electric heater, toaster, electric cooker or kettle.

When an appliance has a metal case, the earth wire is joined to the outer casing, as in Fig. 22.2. The insulation around this wire is of two colours, **yellow and green**, so that even colour blind people can see it is different from the mains supply wires which have one colour each for their insulation. The other end of the earth wire

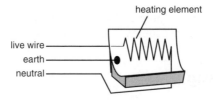

Figure 22.2 The earth wire is connected to the metal case of the heater. A wire protective mesh (not shown) encloses the element

is connected to a very good conductor buried in the ground. We say it is **'earthed'**. The earth connection is often made using a copper water pipe bringing mains water to the building. Now suppose a fault develops in the wiring of an electrical appliance, so that one of the mains supply wires touches the metal case. With no earth connection, the case could give you an electric shock; but with an earth connection in place, a large current flows to earth from the supply, and a fuse in the supply circuit melts. This leaves the appliance isolated from the mains and safe to touch.

Double insulation is a way of providing an extra safety feature. Metal casings of electrical machines have internal seatings made of insulating material. This prevents any contact between the outer casing and the internal metal parts. Better still, the outer casings are often made of insulating materials, so that it is not possible to touch any metal parts.

22.4 Live-neutral-earth

These are the names of the three wires used for mains electricity. In summary:

(a) the **live wire has brown insulation,** and its potential varies from $+340$ V to -340 V relative to earth. **Both the fuse and the on/off switch are connected to the live wire**.

(b) the **neutral wire has blue insulation**. It carries the same current as the live wire and completes the mains circuit.

(c) the **earth wire has yellow and green insulation,** and it connects metal casings to earth. In supply cables the three insulated wires are enclosed in an outer layer of insulation. This provides protection against mechanical damage.

22.5 Residual current circuit breakers

In a mains domestic supply **the live wire is coloured brown, and the neutral wire is coloured blue. The currents in these wires should be the same,** just as the current leaving a cell in a d.c. circuit is the same as the current flowing back to the cell.

If the currents in the live and neutral wires differ by more than a few milliamps, something is wrong. It might be that a small current is flowing through something or someone to earth, but the fuse has not melted. The supply current may not be large enough for that; but it is large enough to give you a shock. The residual current circuit breaker compares the currents in the live and neutral wires; and if they differ by more than, say 20 mA, the circuit breaker activates **a magnetic relay switch called a 'cut out',** and the supply is turned off.

This type of protection has saved lives where people using electric lawn mowers, hedge trimmers, or power tools have accidentally cut through the supply cable with metal parts of their machinery. Nowadays these dangers have been further overcome by using insulators to make the handles and outer casings of many electric tools providing double insulation.

In some homes, the mains supply includes a residual current circuit breaker as well as a fuse box, in order to give people greater protection.

22.6 Wiring a plug

Fig. 22.3 shows the connections inside a typical three-pin plug. The outer insulation of the supply is removed to reveal four or five centimetres of the three separate wires.

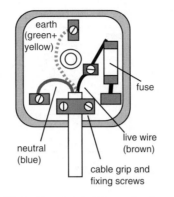

earth
(green+
yellow)

fuse

neutral
(blue)

live wire
(brown)

cable grip and
fixing screws

Figure 22.3 *The view inside a mains three-pin plug*

About one centimetre of the inner insulation must be removed from each of the metal wires before they are connected to the brass contacts inside the plug. The strands of a wire made bare must be kept together, and they are usually twisted together to form a single strand.

The three wires are joined to their contacts by using a small screwdriver to tighten the brass screws. It is important that all the strands of each wire are properly tethered. The brown live wire must be connected to the fuse in the plug which is rated as we discussed in chapter 19.

Before the case is closed, the outer insulation of the cable must be held firmly in the **cable grip,** or under a **strain relief band**, where the cable enters the plug. This prevents tension in the cable from being applied to the separate wires in the plug, which might disconnect them from their contacts (see Fig. 22.3).

22.7 Wiring in the home

Larger currents are carried by thicker wires with thicker insulation. This means that the mains supply wires entering a house must be thick enough to carry the maximum current to be supplied to the house. They first pass through a sealed fuse box. The fuses in the box are to protect the electricity company's supply wires to the building, and the company is responsible for this box.

Fig. 22.4 shows the meter which reveals how much energy has been supplied to the building. The meter readings are taken every three months or so, to cost the amount of energy supplied.

The next installation in a home is the domestic fuse box. Here the circuits are separated for their different uses and suitable fuses for the circuits are installed. An electric cooker may carry the largest current, so thick wires are used here and a 60 A fuse may be installed in the fuse box.

The three pin sockets around the home are wired as in Fig. 22.4 with a 30 A fuse in the fuse box. Added protection is given by fuses of lower rating in each three pin plug, so that the fuse in the fuse box does not melt if a single appliance fails. The supply for the three pin sockets is called a **ring main,** as each of the three supply wires is arranged in a ring. This ensures that where power tools are used, for example, there is no loss of supply voltage at one socket as a result of another socket being in use. The sockets in the ring main are arranged in parallel so that each may be used independently of the others.

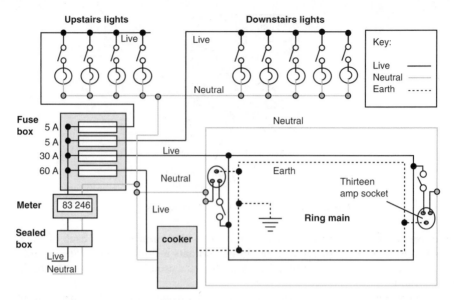

Figure 22.4 *The mains wiring for a house includes switches and fuses in the live wire*

22.8 Lighting circuits

In general, lights use less current than heaters and power tools, so thinner, less expensive wires are used for lighting circuits. There are no earth wires as the cables are concealed in the walls, and metal ceiling light fittings are plastic coated. There are usually two separate lighting circuits, each with a 5 A fuse, so that a single fault cannot put out all the lights at once.

The lights in a house are wired in parallel as in Fig. 22.4, so that they may be used independently. All the circuits we have mentioned have **their fuses and switches in the live part of the supply,** as the live wire is potentially the most dangerous.

As the lighting circuits have thin wires and no earth connections, it is dangerous to connect them to powerful appliances like electric kettles or heaters.

22.9 Lights on the stairs

Fig. 22.5 shows how two-way switches can be used in the live wire to allow a stair light to be turned on or off at two places. The switches are called single pole, double throw (SPDT) switches.

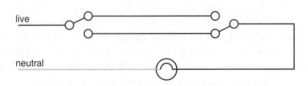

Figure 22.5 *Two double throw (SPDT) switches are used to control a light from two positions*

22.10 Magnetic cut-outs

Magnetic switches may be installed instead of fuses in a domestic fuse box. They operate on the principle that a large current in a coil of wire will magnetise two thin, flexible pieces of soft iron sealed inside the coil. The iron pieces bend when they attract each other, and this movement can be used to turn off the current.

22.11 Costing electricity from the meter readings

We pay for electrical energy according to the number of kilowatt hours (kWh) that have been supplied since the last meter reading. The joule is a very small amount of energy, and the kilowatt hour is much larger. If a power of 1 kW has been used continuously for one hour, then the energy transformed is called **one unit, or one kilowatt hour** of energy.

To calculate the number of units supplied to an appliance we use the relationship:

Energy = power × time.

For example a power of one kilowatt supplied for one hour delivers energy:

= 1 kW × 1 h = 1 kWh.

If we had worked this out in joules, it would have come to:

1000 W × 3600 s = 3 600 000 J,

as 1 kW is the same as 1000 W, and 1 h is the same as 3600 s.

Worked example

Suppose 1 kWh of electrical energy costs 8 p. Calculate the total cost of using the following appliances:

1 a 2 kW washing machine for 1 h,
2 a 1 kW water heater for 1 h and 30 m and,
3 lights of total power 200 W for 5 h.

Answer

	Power (kW)	Time (h)	Energy (kWh)
1.	2	1	2 × 1 = 2
2.	1	1.5	1 × 1.5 = 1.5
3.	0.2	5	0.2 × 5 = 1
			total energy = 4.5 kWh

Total cost = total energy × cost per unit
         = 4.5 kWh × 8 p/kWh = 36 p.

22.12 The future cost of energy

In years ahead, as reserves of fossil fuels like coal and oil become used up, the price of energy may be expected to rise. If energy becomes more expensive, governments might choose to derive more energy from renewable sources like tidal, water wave, solar and wind energy. These sources do not give rise to greenhouse gases, or other forms of chemical pollution, and so they would not be contributing to global warming.

Questions

1   Why does a plastic reading lamp need no earth wire?
2   Explain what is meant when an electricity supply is labelled 100 V a.c. .
3   In Britain the 240 V a.c. supply is also labelled 50 Hz. What do these two numbers mean?
4   Name the wires used for a mains electricity supply, and the colours of their insulations.
5   What two dangers arise if an electric kettle is connected to a lighting circuit?
6   In which wire is a fuse installed in a mains circuit?
7   Why are power sockets and lights each arranged in parallel circuits in the home?
8   When two 60 W, 240 V bulbs are wired in series with a 240 V supply, they glow less brightly than a single 60 W bulb. Why do they not glow like one 120 W bulb?
9   How can a light be switched on or off from two different places?
10 Name a device that can detect and guard against small leakage currents in a circuit.
11 How would you connect an earth wire to a newly-installed electric fire?
12 Imagine you are required to wire up a three-pin plug.
    (a) What colour is the insulation on the wire joined to the fuse?
    (b) Where do you connect the earth wire?
    (c) What colour is the earth wire's insulation?
    (d) Explain why a cable grip is needed inside a three-pin mains plug.

What have you learnt?

1   Can you name the wires used for a mains electricity supply and state their colours?
2   Can you wire up a three-pin plug correctly?
3   Do you know what protects the wiring in the home from risk of overheating?
4   Do you know where the switches and the fuse(s) would be in a.c. circuits?
5   Do you know what is meant by one kilowatt hour of energy?
6   Do you know how to find the cost of using electrical appliances?
7   Do you understand the use of double insulation in a mains cable?
8   Do you know where to connect an earth wire to an appliance?
9   Do you know why electric cooker cables are thick compared with lighting cables when both carry supplies of 240 V a.c.?
10 Do you know why lighting circuits are arranged in parallel?
11 Can you explain the description of an a.c. supply as 240 V, 50 Hz?
12 Do you know what numbers are displayed in a domestic electricity meter?

23 Cathode ray tubes

Objectives

After this chapter you should
- be able to explain how a cathode ray gun works
- know how electron beams may be deflected
- know how a time base circuit is used
- understand how a cathode ray oscilloscope, (CRO), can be used as a timer
- be able to explain both vertical and horizontal deflections in a CRO
- know how an a.c. signal appears when displayed on a CRO.

23.1 Introduction

We learn more about the nature of an electric current in a wire, (see Section 18.3), using the apparatus illustrated in Fig. 23.1. It consists of a sealed glass tube which is highly evacuated. Inside the tube are (i) a tungsten wire joined by two connections to a battery, Y, outside the tube, (ii) a metal plate, P, close to the tungsten wire but not touching it, (iii) a metal plate, Q, a few millimetres away from P and joined to a separate connection outside the tube.

There are two separate circuits. One is a tungsten wire and a battery Y. The second is the metal plate P, connected to the battery X, the ammeter and the metal plate Q.

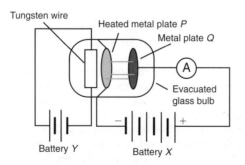

Figure 23.1 'Boiling' electrons off plate P and registering a current through the ammeter

When the battery Y is switched off, no current flows through the ammeter and that is not surprising; there is a gap between the plates P and Q.

However, if battery Y is switched on, the tungsten wire becomes red hot and heats up the metal plate P. Now the ammeter registers a flow of charge.

This suggests that when P is heated up, charge 'boils off' this plate, travels across to plate Q, and continues round the circuit registering a current in the ammeter. If the terminals of battery X are reversed, no current flows in the ammeter. This suggests that **only negative charge is 'boiled off'** plate P and attracted to plate Q, since Q must be connected to the positive terminal of battery X before any current is registered.

More advanced experiments confirm this suggestion. The process of boiling off negative charge from a metal is called **thermionic emission**. It is now known that **the negative charge is carried by particles** which we call **electrons**. **An electric current in a wire is a flow of electrons along the wire**. Metals possess some electrons which are free to move about; and when a metal is heated up enough, some free electrons can 'boil off' its surface into the surrounding space.

23.2 Electron beams

Look again at the apparatus illustrated in Fig. 23.1. When the battery Y is switched on, so that the metal plate P is hot, electrons are boiled off P, and attracted to Q where they are absorbed and continue round the circuit to battery X.

Now imagine what will happen if plate Q has a central hole in it. Electrons leaving P will accelerate towards Q; some will strike Q and some will miss it, and travel through the hole into the space beyond. The stream of electrons emerging from this hole is called **a cathode ray**.

This arrangement is a simple form of **electron gun**, which is found in all television tubes, behind all computer screens and in instruments like cathode ray oscilloscopes. In each instance the gun produces a stream of fast moving electrons, which travel into the space beyond the gun, where they can be deflected.

The section on colour TV in chapter 16 describes the formation of pictures produced when electron beams are deflected inside a cathode ray tube. Two types of deflection are possible. They are discussed next.

23.3 Deflection of electron beams

One way in which an electron beam may be deflected is illustrated in Fig. 23.2. As electrons move through the space between two metal plates which are oppositely charged, they will be attracted towards the positively-charged plate and repelled by the negatively-charged plate. The electrons are deflected from their original path as indicated in Fig. 23.2.

A second means of deflecting an electron beam involves a magnetic field (see Fig. 23.3). Two coils of wire are mounted on the same axis, one on each side of the evacuated glass tube beyond the electron gun. A magnetic field parallel to the axis of the coils results from an electric current in each coil.

When the electron beam reaches this region, it is deflected as indicated in the diagram. This deflection is consistent with the rule for the motor effect (chapter 21), if you remember that a beam of negative charges flowing from left to right is equivalent to a conventional electric current flowing from right to left.

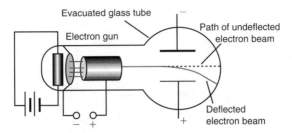

Figure 23.2 The beam of electrons from the gun on the left is deflected in the space between the two charged metal plates

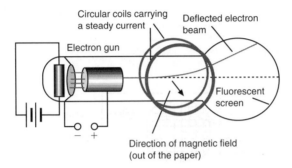

Figure 23.3 Two circular coils carry a steady current outside the tube. The magnetic field between the coils parallel to their axis deflects the beam of electrons upwards

23.4 The cathode ray oscilloscope (CRO)

The CRO has an electron gun which fires a fine beam of fast-moving electrons down an evacuated tube towards a fluorescent screen (see Fig. 23.4). Where the electrons strike the screen, their kinetic energy is transformed into other forms of energy, and a small spot of light is produced. Between the gun and the screen there are two sets of deflection plates.

The main controls of the CRO are indicated in Fig. 23.5. The 'focus' and 'brightness' knobs alter the electron beam passing along the tube towards the screen. The 'focus' adjusts the size of the spot on the screen by controlling the area of the screen that the electrons hit; the 'brightness' alters the number of electrons per second striking the screen, and hence the amount of light produced at the impact spot.

(a) *Measuring a steady voltage*

Imagine that a battery is connected between the two input terminals of the CRO. These terminals are effectively joined to the Y deflection plates (see Fig. 23.4), so that they become charged. If the upper Y plate is connected via the upper input terminal to the positive terminal of the battery, the upper plate becomes positively charged; the lower plate becomes negatively charged.

When electrons pass between the charged plates, they are deflected upwards, being attracted to the upper Y plate and repelled from the lower Y plate. The result seen on the tube's screen is indicated in Fig. 23.6. The upward displacement of the

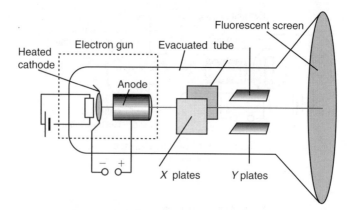

Figure 23.4 In a cathode ray tube the heated cathode emits electrons which are attracted towards the positively charged anode. A fine stream of electrons passes through a hole in the anode and may be deflected by the *X* and *Y* plates before striking the fluorescent screen. A spot of light is produced at the impact point

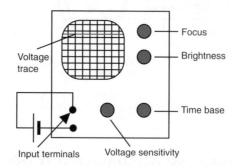

Figure 23.5 A CRO has *Y* input terminals, a tube to display the voltage trace and four main controls

spot on the screen can be measured using a grid of squares ruled on a transparent sheet. This is positioned against the front of the screen.

To find the voltage applied to the *Y* plates, we use the spot's vertical displacement and the voltage sensitivity indicated on the CRO control. So for example, a setting of 5 V/division means that a p.d. of 5 volts applied to the *Y* terminals causes the spot to be deflected by 1 division. Then a deflection of 2 divisions upwards indicates that the battery voltage is 10 V, etc. In short, **vertical displacements on the CRO screen indicate the p.d.s** applied to the input terminals.

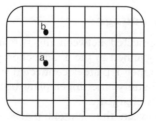

Figure 23.6 Figure (a) shows the spot before the battery was connected. Figure (b) shows the spot after connecting the battery

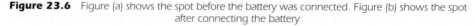

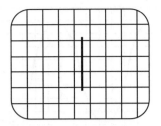

Figure 23.7 The CRO trace is a vertical line when an alternating voltage is applied to the *Y* plates and the time-base is switched off

(b) *Measuring an alternating voltage*

If an alternating voltage is applied to the CRO's input terminals, the spot will move continuously up and down the screen, because the applied voltage is continually changing. If the change in voltage is rapid enough, the spot will move so quickly that it cannot be seen as a spot. It appears as a vertical line on the screen (see Fig. 23.7).

More useful information can be obtained using the CRO time-base (below).

(c) *Using the time-base*

The time-base circuit in an oscilloscope applies a steadily increasing voltage between the two *X* plates in the tube. The result is that the spot on the screen is deflected steadily across the screen horizontally (i.e. in the *X* direction). When the spot reaches the edge of the screen, the time-base returns the *X* voltage to its original value, the spot returns to its starting point, and the cycle is repeated. **Horizontal displacements on the CRO screen represent time intervals**.

The rate at which the spot moves horizontally is indicated by the time-base control knob. Suppose an alternating voltage is applied to the two *Y* terminals, and the trace indicated in Fig. 23.8 is obtained with the time-base setting at 10 ms/division. The trace shows that the voltage goes through a complete cycle in the time that it takes the spot to travel 4 divisions horizontally across the screen. The time for this is 40 ms.

The frequency of the alternating voltage is the number of cycles per second. Using the relation:

frequency (Hz) = 1/time per cycle (s)
                = (1/0.04) Hz = 25 Hz.

This shows how the frequency of an alternating voltage can be measured, as well as its peak-to-peak magnitude. Wave forms may be studied, for example those produced by the sounds from different musical instruments (chapter 14).

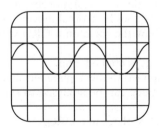

Figure 23.8 With the time-base turned on, the CRO displays an alternating voltage trace

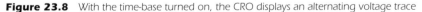

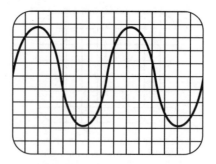

Figure 23.9 A cathode ray tube trace

Worked example
Fig. 23.9 shows an alternating voltage trace on a CRO screen. The voltage sensitivity is set at 2 V/division, and the time-base at 0.5 ms/division. Find:

(a) the peak-to-peak voltage, and
(b) the frequency of the alternating voltage.

Answer
(a) Peak-to-peak voltage = 8 div. × 2 V/div. = 16 V.
(b) Time per cycle = 8 div. × 0.5 ms/div. = 4 ms.

Frequency (Hz) = 1/time per cycle (s)
= (1/0.004) = 250 Hz.

Self test

Use Fig. 23.9 to find (a) the peak-to-peak voltage, and (b) the frequency of the alternating voltage if the voltage sensitivity is 5 V/div., and the time-base set at 1ms/div. (40 V, 125 Hz)

Questions

1   In Fig. 23.1 why is no current detected in the ammeter when battery Y is turned off?
2   In Fig. 23.1 why is no current detected when battery Y is turned on, but battery X is reversed?
3   Draw a diagram of an electron gun and explain how it works.
4   What are cathode rays?
5   List two methods of deflecting cathode rays in a vacuum tube.
6   In a CRO display, what does vertical movement of the spot represent?
7   In a CRO display, what does horizontal movement of the spot indicate?
8   Draw a diagram to show the trace on a CRO when the time-base is turned on and:
    (a) a steady d.c. voltage is applied, and,
    (b) an a.c. signal is applied to the Y plate terminals.
9   Explain how a CRO trace can be analysed to show the frequency of an a.c. supply.
10   Explain the purpose of X plates in a CRO.

What have you learnt?

1 Do you know what is meant by thermionic emission?
2 Do you know what cathode rays are?
3 Do you understand how an electron gun works?
4 Can you explain how charged metal plates can cause cathode rays to be deflected?
5 Can you explain how a current in a coil of wire may cause cathode rays to be deflected?
6 Do you know any experimental evidence that a current in a wire is a flow of negative charges?
7 Do you know how the brightness of the spot on a CRO screen is controlled?
8 Do you know what the vertical and horizontal displacements on a CRO screen represent?
9 Do you know how to use a CRO to measure the frequency of an alternating signal?
10 Can you use a CRO to find the peak-to-peak voltage of an a.c. signal?

⟨24⟩ Generating electricity

Objectives

After this chapter you should
- know that an e.m.f. is induced in a wire if it sweeps across a magnetic field
- know that a rotating armature coil in a motor experiences an e.m.f.
- be able to draw a diagram and explain how an alternating e.m.f. may be induced
- be able to explain the construction of a transformer
- understand step up and step down transformers
- know why the National Grid uses a high voltage supply system
- know how data can be stored on magnetic tape and retrieved.

24.1 Introduction

The motor transforms electrical energy into rotational kinetic energy or work, but the reverse transformation is also possible. The reverse motor effect is called the 'dynamo effect'. **A dynamo transforms rotational kinetic energy into electrical energy**. The dynamo effect may be thought of as an anti-motor effect. The motor causes motion to happen, but the dynamo tends to oppose motion and bring it to a halt.

24.2 The dynamo effect

A conducting wire sweeping across magnetic field lines experiences an e.m.f. along the wire. The generated e.m.f. is larger if the wire is moved faster across the field, or if the field is stronger. The direction of the e.m.f. is reversed if either the direction of the field or the motion of the conductor is reversed. The general name for the production of an e.m.f. using the dynamo effect is **electromagnetic induction**.

There are two laws relating to electromagnetic induction. They deal with the **size** and the **direction** of the e.m.f. induced in a conductor.

Faraday's law: the size of an induced e.m.f. is directly proportional to the rate of cutting magnetic lines of force by a conductor.

Lenz's law: the direction in which an induced e.m.f. acts is so that it tends to oppose the motion or change that caused it.

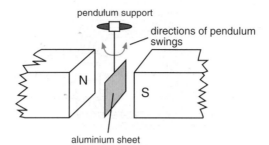

Figure 24.1 *The aluminium pendulum is swung across the magnetic field and soon comes to rest. This illustrates Lenz's law*

To test Lenz's law we could swing a sheet of aluminium in its own plane between the jaws of a permanent magnet (see Fig. 24.1). The metal swings easily as a pendulum, but between the poles of a magnet its swings are quickly damped. To explain the damped oscillations we must consider the induced e.m.f. in the metal when it swings across the magnet's lines of force. The induced e.m.f. causes a current to flow in the metal, and the current flowing in a magnetic field causes a motor effect force on the moving metal. The direction of the force always opposes the motion of the metal in agreement with Lenz's law, so the motion soon stops.

24.3 Dynamos and motors

As soon as the coil of an electric motor (see Section 21.2) begins to rotate, its sides sweep across the magnetic field of its permanent magnet. The motion of the rotating coil in the magnetic field induces an e.m.f. in the coil. As the armature rotations speed up, the induced e.m.f. in the coil increases according to Faraday's law. From Lenz's law, we can tell that this is a back e.m.f., whose direction reduces the current in the coil; **the faster the coil rotates, the smaller is the current supplied to it**. In large industrial motors the starting current, when the coil is at rest, can be so great that it threatens to damage the coil's insulation; and in these motors, as in an electric train for example, it is normal for a protective resistor to be connected in series with the armature. As a train speeds up, the resistance of the protective resistor is reduced, and the **back e.m.f. of the dynamo effect** then prevents the current from becoming too large.

24.4 The a.c. generator

Fig. 24.2 shows a diagram in which a magnet is rotated in the gap in a laminated soft iron ring. A coil of insulated wire is wound on the ring, which can be strongly magnetised by the magnet. As the magnet is rotated, the magnetism in the core changes. When the magnet rotates once, the core's magnetism builds up in one direction and then reduces to zero; then it builds up in the opposite direction and reduces again to zero. **The change in magnetic field directed through the coil generates the**

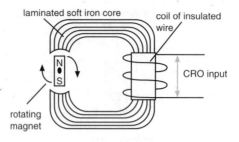

Figure 24.2 An a.c. generator

e.m.f. in it; but this is not a steady e.m.f. Fig. 24.3 shows the appearance of the trace on a CRO (see chapter 23) to which the coil is connected.

The first thing we notice about the CRO trace is that the e.m.f. rises to a peak in one direction before it reduces to zero; then it rises to a peak in the reverse direction and reduces again to zero. This occurs for each rotation of the magnet in the generator. **The number of these cycles of e.m.f. per second is the same as the frequency of rotation of the magnet**. In Britain the mains supply has a frequency of 50 Hz, because the magnets are rotated 50 times each second by steam turbines in the power stations.

The maximum e.m.f. occurs when the magnetic field through the coil is zero. At this point the field through the coil is changing most rapidly and it is the rate of change of the field through the coil which affects the size of the e.m.f.

The current driven by this e.m.f. alternates in direction. The CRO traces show that the peak value of the e.m.f. increases in proportion to the rate of rotation of the magnet. It is greater when a stronger magnet is rotated; and its size is proportional to the number of turns on the coil. So twice as many turns results in twice as large an e.m.f.

Electrical power generating stations apply the principles of electromagnetic induction, whether they are nuclear or fossil fuelled, or hydroelectric power stations (HEPs). All generators have magnets rotated by turbines next to static coils. **It is the relative motion of the magnetic field and the conductors which causes the e.m.f. to be generated**.

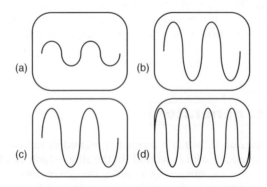

Figure 24.3 (a) A CRO trace shows the a.c. voltage. (b) More volts obtained with a stronger magnet. (c) More volts obtained with more turns on the coil. (d) More volts and more cycles per second when the magnet is rotated faster than in (a)

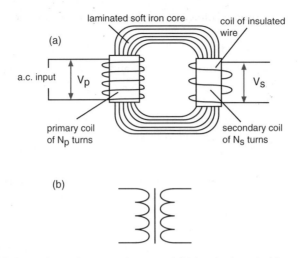

Figure 24.4 (a) A step-down transformer and (b) the circuit symbol for a transformer

24.5 Transformers

The transformer is like the a.c. generator shown in Fig. 24.2, but the gap with the rotating magnet has been replaced as in Fig. 24.4 with another insulated coil wound on the soft iron ring. Alternating current flowing in this new coil magnetises the core just as it was magnetised by a rotating magnet. As a result, an alternating e.m.f. is induced in the second coil just as if it was the output coil of an a.c. generator.

Laminated soft iron describes the materials used for the armatures of motors and cores of transformers. Thin strips of soft iron are dipped in varnish and then bonded together. The varnish **prevents eddy currents** from flowing from one strip of soft iron to another when magnetic fields in the core are varying. So the iron does not get very hot, and transformers become more highly efficient.

At first sight the transformer seems to have achieved no advantage since a.c. in its **primary coil** sets up a.c. in its **secondary coil**. However, the two coils are electrically separate, being isolated from each other and from the core material. None of the electric charges flowing in the primary coil flow into the secondary coil, and this is sometimes an advantage. What links the coils very effectively is the magnetic field induced in the iron core, and **only when the magnetism in the core changes** can e.m.f.s be induced in the coils. It follows that a steady direct current in the primary coil induces no e.m.f. in the secondary coil, since there is no change in magnetism linking through it. Transformers do not work with direct currents, d.c., which flow from cells and batteries.

The chief advantage of a transformer can be seen when we use primary and secondary coils with different numbers of turns on them. A double beam CRO can display traces representing the p.d.s across the two coils at the same time. Comparison of the traces shows that the primary and secondary voltages are in the same ratio as the turns ratio for the primary and secondary coils. We can step the voltage up to a higher value if there are more turns on the secondary coil than on the primary coil. Similarly, we can step the voltage down if there are fewer turns on the secondary coil than on the primary coil. The turns ratio equation states:

secondary voltage/primary voltage = number of secondary turns/number of primary turns

In symbols: Vs/Vp = Ns/Np, (24.1)

where N means the number of turns on a coil, and s and p refer to the secondary and primary coils respectively.

Self test

1 A mains supply of 240 V a.c. has to be stepped down to a supply of 6 V a.c. A transformer with 400 turns on its primary coil is available. How many turns are needed on its secondary coil? (10 turns)
2 If the transformer in the previous question had the same primary coil and a secondary coil of 600 turns, what would be the size of the secondary voltage? (360 V a.c.)

By stepping the secondary voltage up in a transformer, we seem to be gaining something for nothing. However, in terms of energy or power there is no gain. Transformers can be very efficient, so that perhaps more than 95% of the energy supplied to them becomes useful energy output. This means a transformer can work without becoming extremely hot.

A good approximate calculation can be made assuming that a transformer is 100% efficient. In words we are assuming:

power input to the primary coil = power output from the secondary coil

In symbols: V_p(volt) $\times$ I_p (amp) = V_s (volt) $\times$ I_s (amp) (24.2)

24.6 The National Grid system

Fig. 24.5 illustrates the distribution of electrical power from a power station to a domestic user. A step-up transformer is used at the power station, so that the p.d.

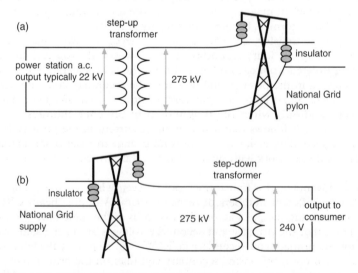

Figure 24.5 (a) Transformers step up the voltage of the supply at the power station. (b) Near the consumer the voltage is stepped down to 240 V

between the cables supported on pylons across the country is raised to 275 kV, or 400 kV. These cables are part of the National Grid system. At the consumer's end of the cables the voltage is stepped down through a series of transformers, so that the p.d. of the mains supply to private houses is only 240 V.

The efficiency of electrical power distribution depends largely on the amount of power lost in heating the supply cables of the National Grid. A current in a resistor causes heating, and we saw in chapter 19 that the rate of heating, or power loss, could be calculated from the equation:

Power loss = (current)^2 × resistance   (equation 19.10)

This power is wasted in heating the cables of the National Grid, so there is an advantage in keeping both the current and the resistance of the cables as small as possible.

Equation 24.2 shows us that, for a given power transfer, the current required is smaller if the a.c. voltage is made greater. **The high voltage of the National Grid allows the current in the cables to be kept low**, and the cables themselves are made thick enough to keep their resistance low. This ensures that the a.c. supply via the National Grid can be delivered with much higher efficiency (i.e. with much less energy loss), than would be possible if the supply voltage was much lower.

The advantages of alternating currents in power distribution include:

(a) it is easier to generate a.c. than d.c. using turbine generators in power stations
(b) it is easier to transform a.c. to higher and lower voltages than to transform d.c. voltages
(c) less power is wasted when the supply voltage is raised by a step-up transformer at the power station and lowered by a step-down transformer near the consumer.

24.7 Dangers of high voltages

As the p.d. across a conductor rises, the current through it increases. High voltages can be damaging to apparatus and cause harmful shocks to a person's nervous system. An electric shock occurs when tissues in the body carry electric currents from sources outside the body. We can experience life-threatening shocks from the mains a.c. supply of 240 V, especially in damp conditions.

Near a pointed conductor at a high voltage there is an electric field which can strip electrons from neutral air molecules. This ionising process makes the air a conductor, and a spark may jump from one pointed conductor to another.

Electricity pylons carrying the National Grid supply support the cables on specially designed insulators, so that the pylons are insulated from the supply voltage. Leakage from the cables is possible in damp atmospheres which conduct electricity better than dry ones. So the cables must be supported far enough apart and high enough above the ground, to prevent large leakage currents.

24.8 Magnetism and data storage

The development of magnetic tape provided a ready means of recording signals and retrieving them later. The tape carries a thin layer of a powdered magnetic material. When it is run close to unlike poles of a magnet, the magnetic layer on the tape becomes magnetised with unlike poles at opposite edges of the tape. If the magnetic field is reversed, the tape is magnetised in the opposite direction.

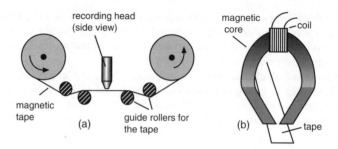

Figure 24.6 (a) Tape is guided past a recording head in a tape recorder. (b) The recording head has a gap across which magnetic fields magnetise the tape from edge to edge

Now suppose the magnetic field is caused by a **recording head** and not by a permanent magnet. The recording head is an electromagnet with a narrow gap where the magnetic field is concentrated (see Fig. 24.6). As the tape passes this gap, its induced magnetism depends on the currents in the electromagnet. So the currents which can energise a loudspeaker to provide music can also magnetise the tape in a tape recorder, and the tape can store up the data contained in the musical sounds. The magnetism in the tape matches the variations in atmospheric pressure associated with the sound waves it records. Data stored in this way is said to be in **analogue form**.

When we wish to 'play' the tape and hear the music, we run the tape again past the gap in the recording head at the same speed as during the recording. The magnetism of the tape induces e.m.f.s in a coil on the recording head. This coil is connected to the input terminals of an amplifier from which a loudspeaker is energised. So playback involves retrieving data from the tape by electromagnetic induction.

Self test

If the tape were played back faster than its speed during recording, how would the pitch of sounds be affected?

(Higher frequency signals means higher pitch).

Data can also be stored on the tape in **digital form**. Since the tape can be magnetised in two opposite directions, it can store binary numbers. These numbers consist entirely of ones and zeros. So information **encoded** into binary numbers can be stored and retrieved as before. Binary signals have the advantage that they can store information more accurately than analogue signals, but extra circuits are needed to encode and decode the signals.

Questions

1   What causes an induced e.m.f. in the armature coil of an electric motor?
2   How is the induced e.m.f. in Q.1 affected when the armature rotates faster?
3   (a) What happens to the current supplied via the brushes to the armature coil of an electric motor, when the armature rotates faster?
    (b) Whose law summarises the effect you have described in part (a)?
4   (a) Draw and label a diagram to show the parts of a simple a.c. generator.
    (b) What controls the frequency of the induced e.m.f. in the generator?
    (c) List two ways of inducing a larger alternating e.m.f. at a fixed frequency.

5 (a) Why is the voltage of the National Grid system very high?
  (b) How is the high voltage for the National Grid obtained?
  (c) How is the 240 V a.c. supply for domestic use obtained from the grid system?
  (d) What dangers exist for people near conductors at very high voltages?
6 (a) Power of 24 kW is supplied by an a.c. supply of 24 kV. What current flows?
  (b) The same power is supplied by an a.c. supply of 240 V. What current flows?
  (c) If the power cables in (a) and (b) above each have resistance of 1 ohm, how much power is used heating the cables in (a), and in (b)?
  (d) Show which distribution system in (a) and (b) above is the more efficient.
7 (a) Draw and label the parts of a transformer.
  (b) Explain how a transformer can double an a.c. voltage.
  (c) If the transformer in (b) is 100% efficient, compare its output and input currents.

— **What have you learnt?** —————————————————————————

1 Can you describe both Faraday's law and Lenz's law of electromagnetic induction?
2 Can you explain how the rotating coil in a motor generates its own e.m.f.?
3 Can you explain why electromagnetic induction is called a reverse motor effect?
4 Can you draw a diagram to explain how an a.c. generator works?
5 Do you know the action of a transformer?
6 Can you write down the turns-ratio equation for a transformer?
7 Can you write down the power equation for a transformer which is 100% efficient?
8 Do you know how transformers are used in distributing power in the National grid system?
9 Can you describe the dangers threatened by conductors at high voltages?

25 Electronics in control

___ **Objectives** _____

After this chapter you should
- understand how transistors can be used as switches
- understand that logic gates can control transistor switches
- understand how magnetic relay switches can be used to operate powerful devices
- understand the uses of input control, processor, and output transducer in electronics
- be familiar with truth tables for logic gates
- understand how potential dividers are used for input controls to logic gates
- know how a bistable circuit behaves
- know what components are used as sensors and for time delays.

25.1 Transistor switches

A transistor is a device made of three layers of semiconductor materials arranged like a thin sandwich as in Fig. 25.1a. The layers are called **collector** (c), **base** (b), and **emitter** (e), and their conducting properties depend on small amounts of impurities they contain. There is a separate electrical connection to each layer. Fig. 25.1b shows the circuit symbol representing a typical transistor, and Fig. 25.2 shows a circuit diagram including a transistor used as a switch. Below the circuit a block diagram shows the circuit's logic. There is **a light sensor, a transistor switch** and **a bulb**, which are particular examples of **an input sensor, a processor** and **an output transducer** respectively.

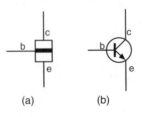

(a)   (b)

Figure 25.1   (a) A transistor has three layers called base, b, collector, c, and emitter, e. (b) The circuit symbol

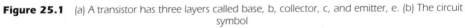

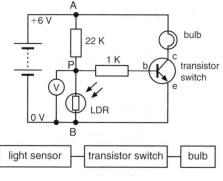

Figure 25.2

In Fig. 25.2 you should identify the symbols for a light bulb, two resistors, a light dependent resistor (LDR), a 6 V battery and a solid state or high resistance voltmeter. When you connect the circuit in Fig. 25.2, the bulb does not light up as current is unable to flow through the transistor from the collector to the emitter. The voltmeter reading is low, but it increases if you put a light-proof cover over the LDR. When the voltage across the LDR increases, the light bulb is turned on. A certain voltage between the base and emitter of the transistor is needed to inject a small current into the base, and this enables a larger current to flow from the collector to the emitter. What happens when you uncover the LDR? Can you explain why the bulb goes out?

The 1 K (1000 ohm) resistor connected to the base protects the transistor from damage caused by a large base current. This resistor ensures that only a few milliamps can flow into the base from the 6 V supply; but the resulting collector current can be as much as sixty times bigger, large enough to light up the bulb. The transistor switch is being used as a **current amplifier**, but it is only **turned on if the voltage drop from the base to the emitter is large enough**.

25.2 Controlling the transistor

The resistance of the LDR in the dark is possibly as high as 100 K. Then, if it is in series with the 22 K resistor, the LDR has the majority of the resistance between A and B in Fig. 25.2; so most of the supply voltage will be across the LDR. The voltage-drop from P to B will be high, and so the transistor will be 'turned on'.

In a well-lit room, the resistance of the LDR is lower, possibly as low as 1 K. If it is now in series with the 22 K resistor, the LDR has only a small fraction of the total resistance between A and B, and only a small fraction of the supply voltage is maintained across it. This means the voltage-drop from P to B is low and the transistor is 'turned off'.

By changing the size of the series resistor, we can change the light level at which the bulb lights up. Try 10 K in place of the 22 K series resistor. The LDR should then need a proportionally smaller resistance to raise the voltage at P enough to switch on the transistor. The circuit should, therefore, be more sensitive to failing light. On the other hand, if you try using a 47 K series resistor, the LDR will have to be in darker conditions before the bulb is lit. To make this circuit more generally useful, a variable resistor would enable the user to select the level of failing light which would switch on the transistor and light the bulb.

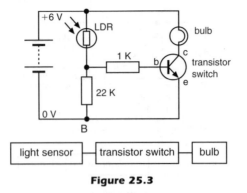

Figure 25.3

Fig. 25.3 shows a similar circuit in which the LDR and its series resistor have been swapped. They still act as a potential divider to control the transistor, but what must be done to turn on the bulb? This time the bulb is lit in bright light, and goes out when the LDR is in the dark. So in different circuits the LDR can be used either to switch on a transistor as it gets light, or as it gets dark.

The logical sequence of actions in electronic circuits may often be represented by a block diagram:

Input control ➤ **Processor** ➤ **Output transducer**

Fig. 25.4 shows a typical circuit assembled on a printed circuit board, or pcb. The components of a temperature monitor are shown neatly arranged on top of the board, and their terminals pass through small holes drilled in the pcb. The terminals are then soldered to copper strips bonded to the underside of the board so that all the circuit connections are completed. Notice the rows of sausage-shaped resistors, a larger

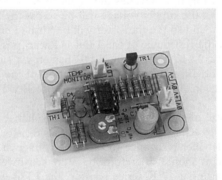

Figure 25.4 An electronic temperature monitor circuit assembled on a printed circuit board (Courtesy of Projects and Modules Team, Maplin Electronics plc)

adjustable (or preset) resistor of up to 10 K at the near edge, the dark **processor** mounted centrally, and some capacitors C1, C3 and C4. The **input control** connection TH1 from a thermistor is on the left, while the 12 V d.c. power supply is connected at the other end of the board. The output terminal from the pcb to an **output transducer** (e.g. a relay switch not shown) is near the centre of the far edge of the board.

The next three sections describe the stages mentioned above in the block diagram.

25.3 Input controls

Potential divider controls are widely used where one of the two resistors is variable. For example, the resistance of a **thermistor** falls as it warms up. In Fig. 25.5 it is used in a frost-warning device. As it gets cold, the resistance of the thermistor increases,

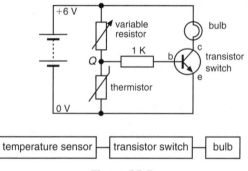

Figure 25.5

which means that the voltage at Q increases. The transistor is then turned on and the bulb is lit. The variable series resistor can be adjusted so that the temperature at which the transistor switches can be altered.

Thermistors are also used in televisions and computers to prevent sudden surges of electric current when you switch circuits on. At room temperature the thermistor has a high resistance which allows only a small current to flow, but gradually it warms up, its resistance falls, and the current increases; so the thermistor enables the full working current to be turned on gradually.

--- Self test ---

Can you change Fig. 25.5 so that the bulb is lit as the air temperature rises?

A moisture sensor has a high resistance when it is dry, but a lower resistance in the rain. It consists of narrow strips of bare metal fixed to an insulating support as in Fig. 25.6. In order to turn on the transistor in Fig. 25.6, the moisture detector must have low resistance, so it must be wet. This could act as a rain warning circuit, or be used inside a water tank to show when the tank is full.

Swapped with its series resistor in Fig. 25.6, the moisture detector could switch on the transistor in dry conditions, so it might show when a pot plant needs water, for example.

A pressure pad is also a sensor. It has wire contacts with two thin sheets of metal

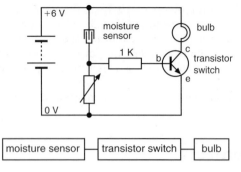

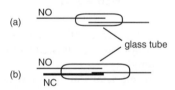

Figure 25.6

inside a protective cover. The sheets are held apart by a layer of spongy insulating material with holes in it, and contact is made between the sheets of metal when you tread on the pad. The resistance of the pressure pad is very small when the metal sheets are in contact.

Self test

Design a circuit to turn on a bulb when someone treads on a pressure pad.

The next sensor is sometimes used in doorways instead of a pressure pad under the mat.

A reed switch closes or opens when a magnet is laid alongside it. Switches as in Fig. 25.7a are normally open (NO), but normally closed (NC) switches are also

Figure 25.7 Figure (a) shows a normally open reed switch (NO). Figure (b) shows normally open and normally closed contacts. The thicker line indicates the non-magnetic contact

obtainable. The resistance of the switch is very low when its contacts touch. A magnet near two flexible iron strips inside a glass tube can induce magnetism in the strips so that they attract each other and make contact. This occurs in the normally open switch in Fig. 25.7a.

In Fig. 25.7b there are two metal strips on the left end, one of which is not a magnetic material. When the magnet is nearby, the magnetic strips make contact and the non-magnetic strip is isolated. With no magnet nearby, the magnetic strips lose contact and the non-magnetic strip is connected again as in a single pole double throw switch (SPDT).

A shopkeeper may have a pressure pad under the door mat, or a reed switch set in the door frame and a small magnet on the door. These arrangements are used as sensors either to show when someone treads on the mat or opens the door, respectively. In Fig. 25.8 the transistor switch will light the bulb only while the pressure pad is compressed.

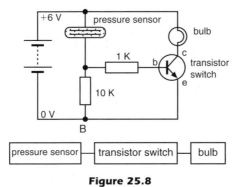

Figure 25.8

A summary of input controls might include the following list of sensors:

Input control components	Relating to
LDRs	Light intensity
Thermistors	Temperature
Moisture detectors	Surface moisture
Pressure pads	Applied pressure
Reed switches	Magnetic effects
Infra-red detectors	Long wavelength light
Microphones	Audible sound
Ultrasonic detectors	High frequency sound
Timing capacitor	Time delays (see later)

25.4 Output transducers

Output transducers are devices which can be switched on and off by a transistor. The example we have used is the **light bulb**; but a shopkeeper might prefer to hear a **buzzer** when the door to the shop is opened. A small buzzer could replace the bulb as an output transducer. The same could also be said of a red, green or yellow **LED**, whose current can be supplied through a protective series resistor.

Suppose we want to turn on street lights when it gets dark. The lights require another power source and more current than many transistors can supply. They cannot be operated by connecting them to the collector of the transistor. In another example we may want to close the curtains automatically when it gets dark. The electric motor we need takes more current than the transistor can supply. To solve these problems we need another kind of switch (or another kind of transistor).

Fig. 25.9 includes a **magnetic relay switch** or **reed relay**. A reed relay has a reed switch set inside a coil of insulated wire. A current in the coil magnetises the reed switch, so that its contacts close. This turns on the power supply in a second circuit. When the transistor is turned off, the magnetic relay switch is opened. The advantage of a relay switch is that a large current can be turned on by means of a smaller one; and even a mains alternating current (a.c.) can be controlled by means of the small direct current (d.c.) in the transistor. A great many street lights, garage doors, security locks and electric motors are controlled by means of relay switches.

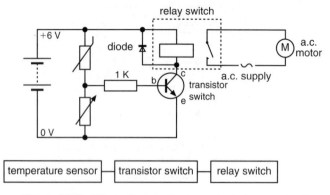

Figure 25.9 *What will cause the a.c. motor to be energised?*

A diode is always connected as shown in Fig. 25.9, whenever a magnetic relay switch is used. Placed as shown in the figure, it does not pass current to the collector from the 6 V supply. But when the current in the relay's coil is turned off, a sudden large voltage 'spike' may be generated in the coil for a fraction of a second. This induced voltage could damage the transistor, but the effect is prevented by the diode which provides a closed loop with the coil and dissipates its energy harmlessly in the coil itself.

Block diagrams are used to illustrate the stages used in the design of apparatus. For example:

| Smoke detector | ➤ | Processor | ➤ | Sound emitter |

This diagram shows the logical sequence of stages in a smoke alarm system.

___ **Self test** _____

Draw block diagrams to show the stages needed for the following circuits:

1 You want to use an electric motor to open greenhouse windows when the air warms up.
2 You want to sound a buzzer when someone opens a door.
3 You want to light a bulb when someone sits in a car seat.

A summary of output transducers might include the following components:

Output transducer	Releases what form of energy?
Light bulb	Heat and light
LED	Light
Buzzer	Sound
Bell	Sound
Relay and secondary circuit transducer	Various

25.5 Processors

At the centre of an electronic device is a **processor**. The transistor switch is a good example, as it receives input voltages and acts in response to them. Its actions cause output transducers to be turned on and off.

The processor could be an **electronic calculator** which responds to input voltages it receives when you press the pads. It might even be the **central processing unit** (CPU) of a computer. These are examples of processors in which a very large number of transistor switches are used to process the input signals. Their output transducers may be liquid crystal displays, computer screens, printers or loudspeakers in a multimedia system.

Logic gates are processors which use fewer transistors. As in more complex processors the outputs of logic gates are either high voltage (logic 1), or low voltage (logic 0). The gates can drive only small currents from their outputs. They may, for example, light LEDs but not more powerful lamps. Power transistors may be controlled by means of logic gates to drive larger currents.

Processors include transistor switches, logic gates and computer processors (CPUs). Of these, a transistor normally operates when a current flows into its base; but logic gates by comparison can take negligible input current and behave purely as voltage switches.

25.6 More about logic gates

When logic gates are used as electronic switches, the output from one gate can be an input to a second gate, so that a sequence of logic processes may be carried out.

The symbols for two logic gates are shown in Fig. 25.10. The d.c. amplifier gives a

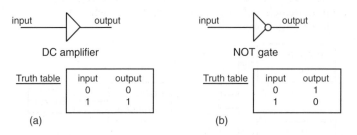

Figure 25.10 (a) The DC amplifier converts various voltage waveforms into square waves, and it can supply a small output current to light an LED. (b) The NOT gate is an inverter and has a circle on its circuit symbol

high output voltage when the input voltage exceeds the switching voltage; otherwise, its output voltage is low (logic 0). This behaviour is shown in the **truth table** beside the circuit symbol.

The second gate shown is a NOT gate, the output from which is NOT the same as its input. The truth table for this is also shown beside the circuit symbol. Notice the symbol has **a small circle** on its output side. **Inverting gate symbols in general have this circle**.

The other gates each have two inputs and one output. Fig. 25.11a shows an AND gate, the output from which is high (logic 1), when one input AND the other are both

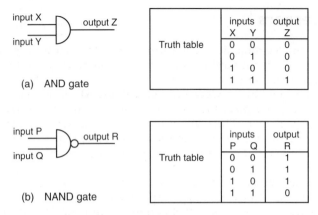

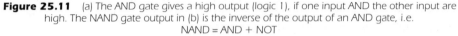

	inputs X Y	output Z
Truth table	0 0	0
	0 1	0
	1 0	0
	1 1	1

(a) AND gate

	inputs P Q	output R
Truth table	0 0	1
	0 1	1
	1 0	1
	1 1	0

(b) NAND gate

Figure 25.11   (a) The AND gate gives a high output (logic 1), if one input AND the other input are high. The NAND gate output in (b) is the inverse of the output of an AND gate, i.e.
NAND = AND + NOT

high. The NAND gate in Fig. 25.11b gives the inverse of this. i.e the output is NOT the AND output.

Fig. 25.12a shows an OR gate symbol and truth table. The output is high if one input OR the other, OR both inputs, are high. And the NOR gate gives the output which is NOT the OR output.

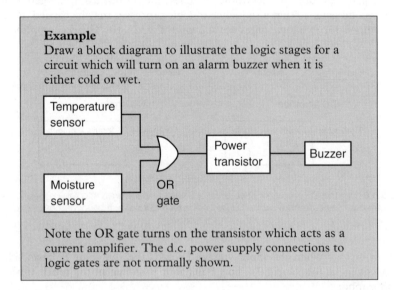

Example
Draw a block diagram to illustrate the logic stages for a circuit which will turn on an alarm buzzer when it is either cold or wet.

Note the OR gate turns on the transistor which acts as a current amplifier. The d.c. power supply connections to logic gates are not normally shown.

Self test

1   Write truth tables for each of the combinations in Fig. 25.13.
2   Which single logic gate has the same truth table as in these examples?

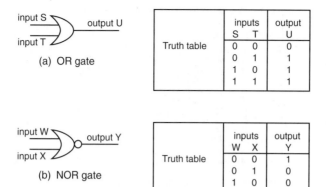

(a) OR gate

		inputs S T	output U
Truth table		0 0	0
		0 1	1
		1 0	1
		1 1	1

(b) NOR gate

		inputs W X	output Y
Truth table		0 0	1
		0 1	0
		1 0	0
		1 1	0

Figure 25.12 (a) An OR gate gives high output if one input OR the other OR both inputs are high. In (b) the NOR gate gives a high output if neither one input NOR the other is high, i.e. NOR = OR + NOT

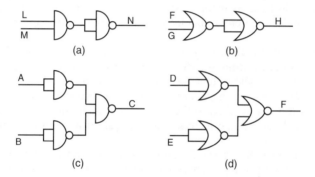

Figure 25.13 Arrange truth tables for these arrangements, and find a single gate for each which behaves in the same way

25.7 A bistable circuit

A circuit can be connected using two NAND gates (see Fig. 25.14). Notice that the output of each gate is connected to an input of the other gate. Two other input connections, called **set** and **reset**, are each held at logic 1; but they can be taken to logic 0 by means of push switches. When one of the two outputs is at logic 0, the other is at logic 1. A bistable unit can also be made from two NOR gates with set and reset inputs held at logic 0.

When switch A in Fig. 25.14 is pressed, the set input goes to logic 0, and the output goes to logic 1, which turns on the transistor switch. The buzzer sounds continuously even when switch A is released, or pressed a second time. The output is said to be **latched**. This is useful for a security system which sounds an alarm continuously once it has been triggered. The concealed switch B is used to turn off the alarm and reset the bistable to its original state. This circuit, known as a **bistable, latch** or **flip-flop** is also used as a memory device which stores a zero or a one.

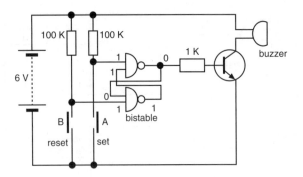

Figure 25.14 The bistable circuit latches so that, once triggered by switch A, it sounds the buzzer until it is reset by switch B

25.8 Time delays

A time delay can be introduced so that a logic gate is 'triggered' after a fixed interval. An egg timer, for example, is turned on when the egg is put into boiling water. Nearly four minutes later an LED lights up to show the boiled egg is ready to eat. The timing input control includes a potential divider with a variable resistor and a **capacitor** (see Fig. 25.15).

When the set switch is closed, the battery is connected to the rest of the circuit and **the capacitor begins to store up the electric charge** flowing through the variable resistor. A larger resistor allows less current to flow, so it takes longer to build up the charge in the capacitor. As the charge reaches the capacitor, it raises the voltage developed across this component, so the voltage at X increases. After a certain interval the voltage at X is large enough to trigger a logic gate from whose output the LED is lit.

When the voltage at X triggers the NOT gate, the output at Z switches to logic 0; so there is then a drop in voltage between Y and Z. Current then flows in the LED and it lights up. Note that the current to the LED is small as it has a series protective resistor.

The length of the time delay increases with the resistance, R, and also with the capacitance, C, of the capacitor. Capacitance is measured in farads, (F), or microfarads (μF). The time delay is best set by trial and error using a variable resistor of 2.2 megohms and a capacitor of 220 μF. This type of **capacitor has its positive and negative terminals labelled**, and it must be connected as shown.

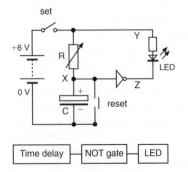

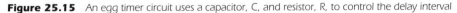

Figure 25.15 An egg timer circuit uses a capacitor, C, and resistor, R, to control the delay interval

Time delays are useful in security circuits. Once a burglar alarm with a bistable has been set and latched, a time delay allows you to reset the device before it goes off. That way it need not go off every time you open your front door. Time delays of a fraction of a second are also possible if the selected resistance and capacitance values are small.

Capacitors with small capacitance may safely be connected either way round in a circuit, provided their working voltage is not exceeded.

25.9 Converting a.c. to d.c.

In Fig. 25.16 an alternating supply is shown in series with a diode and a resistor. A double beam CRO displays both the a.c. signal and the voltage across the resistor. Notice that current flows through the resistor in only one half of each a.c. cycle, because the diode blocks current in the reverse direction. The process of producing these d.c. voltage pulses is called **half wave rectification**.

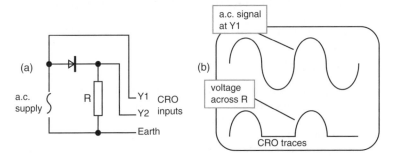

Figure 25.16   (a) Half wave rectification of an a.c. supply by means of a diode. (b) The rectified signal shows half of each input cycle of a.c. is blocked

Full wave rectification can be achieved when four diodes are connected as shown in Fig. 25.17a. In one half of the a.c. cycle, current flows from A to D along the path ABCD, giving rise to a voltage pulse across the resistor with B at the more positive voltage. In the second half of the a.c. cycle, current flows from D to A along the path DBCA, giving rise to another voltage pulse across the resistor in the same sense as before. Fig. 25.17b shows the shape of the CRO display of the voltage across BC.

The d.c. voltages produced by diode rectifiers are pulses rather than steady d.c. voltages. They can be smoothed into much more constant voltages by means of large capacitors, called smoothing capacitors, connected in parallel with the resistors shown in these figures

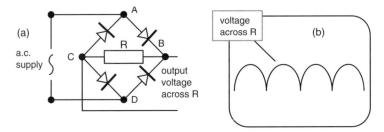

Figure 25.17   (a) Full wave rectification by means of four diodes. (b) The CRO shows a d.c. voltage pulse for each half cycle of the a.c. supply

1 Name two input sensors and two output transducers used with transistor switches.
2 What type of circuit can control a transistor used as an electronic switch?
3 How is an a.c. supply turned on using a transistor switch when it gets dark?
4 Draw a circuit diagram to show an AND gate giving a high output in damp, warm conditions.
5 Draw a block diagram to show a transistor and lamp turned on in warm AND damp conditions.
6 Draw a block diagram to show how a buzzer can be turned on in warm OR dry conditions.
7 Draw a circuit diagram to show how a magnetic relay can turn on an a.c. supply to a heater when conditions turn cold.
8 Draw a bistable circuit diagram with outputs which turn on a red LED when it is set, and a green LED when it is reset.
9 Give examples of two types of processor used in electronics.
10 (a) Draw a potential divider suitable for causing a time delay in a switching circuit.
   (b) How could you reduce the length of the delay interval?

What have you learnt?

1 Do you know what are the base, collector and emitter of a transistor?
2 Do you know which component is a temperature sensor?
3 Can you draw a control sensor circuit to produce a high voltage connection (logic 1) in the dark?
4 Can you produce a truth table for a two input (a) NAND gate, and (b) NOR gate?
5 Do you understand how to set and reset a bistable circuit?
6 Can you draw block diagrams to show the logic stages in electronic switching circuits?
7 Do you understand the need for a diode in a circuit containing a magnetic relay?
8 Can you produce a circuit using just NAND gates, so that the circuit behaves as a NOR gate?
9 Do you understand how a capacitor can be used to cause a time delay?

26 The Earth and the solar system

Objectives

After reading this chapter you should
- be able to describe the Solar system
- understand the motions of the Earth and the planets
- understand seasonal changes
- understand the changes in the appearance of the Moon
- understand eclipses of the Sun and Moon
- understand that the Moon effects the tides on Earth
- understand that conditions on the surfaces of other planets differ from those on Earth
- understand that some comets orbit the Sun in long, thin, elliptical orbits.

26.1 Introduction

The study of the Sun, the stars and the planets, called astronomy, is the oldest branch of science, going back at least as far as the beginnings of recorded history. To primitive man, the Earth was the centre of the universe. Gradually this view has changed, and no point can be identified as more central than another.

Observations show that **the Earth and other planets orbit the Sun** in what is known as **the solar system.** Our Sun, together with its solar system of planets and all the other stars we can see with the naked eye, and more besides (about one hundred thousand million of them), make up our star system or **galaxy,** sometimes called the **Milky Way.** Other, more distant galaxies (see Fig. 26.1), can be detected with large telescopes.

Our Sun is an average sort of star. Some other stars are much bigger and much brighter than the Sun; but the Sun seems more important and much larger than other stars because it is so much closer to us. Its distance from the Earth is about 150 million kilometres; sunlight takes about eight minutes to reach the Earth. For more about distances to other stars see Section 27.1.

26.2 Light in the solar system

The Sun is a brilliant source of light. It is the only self-luminous body in the solar system. We can see other bright objects like the Moon and the planets, because some of

Figure 26.1 A galaxy contains many millions of stars. This galaxy, known as M 81, is found in the constellation Ursa Major
(Mount Wilson and Palomar Observatories)

the sunlight reaching them is then scattered in our direction. If sunlight ceased to fall on them, these objects would become invisible (see Fig. 26.2).

The stars are also sources of light on their own. The source of energy in the stars is discussed in Section 27.3.

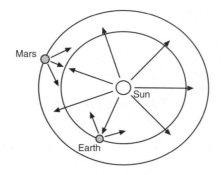

Figure 26.2 Mars emits no light of its own. We only see the sunlight that it scatters in our direction

26.3 The seasons

The Earth has two different motions at the same time. First, it is spinning or rotating once every 24 hours (see Fig. 26.3). Its axis of rotation passes through the two geographical poles (North and South), which mark the positions of latitudes 90 degrees N and 90 degrees S respectively.

The half of the Earth facing the Sun at any moment is light; the other half is dark. The rotation of the Earth means that most places experience a period of light and a

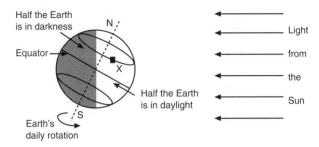
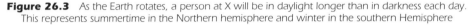

Figure 26.3 *As the Earth rotates, a person at X will be in daylight longer than in darkness each day. This represents summertime in the Northern hemisphere and winter in the southern Hemisphere*

period of dark every 24 hours. Once every 24 hours we see the Sun rise in the East and set in the West and then, after a spell of darkness, the sequence repeats itself.

The second motion of the Earth is a gradual movement round the Sun. Its complete passage or orbit takes one year (see Fig. 26.4). During this time, the axis of the Earth's daily rotations remains pointing in the same direction in space, with the axis beyond the North pole pointing up approximately towards the pole star. The two motions of the Earth give rise to the sequence of seasons.

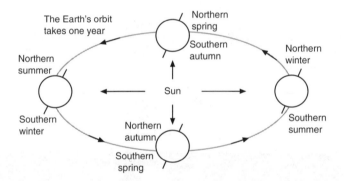

Figure 26.4 *The seasons change as the Earth orbits the Sun. The Earth's axis of rotation points in the same direction throughout the year, giving long hours of daylight in summer and darkness in winter*

26.4 The Moon

The Moon appears from Earth to be about the same size as the Sun, but it is much smaller than the Sun and very much closer to us. It is our nearest neighbour in space, at a distance of about 380 000 kilometres. The Moon is smaller than the Earth, and the gravity experienced at its surface is much less than on the Earth, so that the Moon can retain no gaseous atmosphere. It orbits the Earth in a period of about one month, and spins on its axis at the same time. This means that it always has the same face directed towards us (see Fig. 26.5).

We can only see the Moon because half of its surface is illuminated by sunlight, some of which is scattered in our direction. Seen from the Earth, the Moon goes through a cycle of phases illustrated in Fig. 26.6.

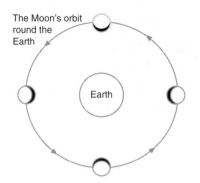

Figure 26.5 *Four positions of the Moon are shown as it orbits the Earth. The Moon rotates so that the same part of its surface is always towards the Earth*

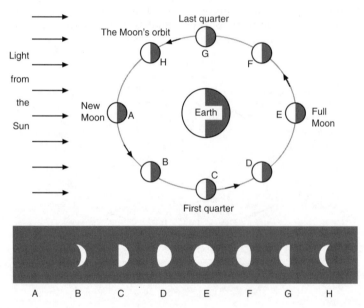

Figure 26.6 *The lower diagram shows how the Moon appears on Earth during each orbit which lasts about 29 days. The letters A–H below relate to the positions of the Moon in the top diagram*

As the Moon rotates, a point on its surface may be in continual sunlight for about a fortnight, and then continually in the dark for another fortnight. The surface at this point experiences a large range of temperature, from 130°C at its hottest down to −170°C at the end of the dark phase. You can understand why the time and place of the Moon landing by astronauts needed careful planning. The Moon's surface temperature needed to be close to the human body temperature (37°C), or else the astronauts would have had to carry complex heating or cooling systems in order to survive.

Figure 26.7 The first manned landing on the Moon. Aldrin steps onto the Moon's surface (NASA)

The Apollo Mission of July, 1969, was the first in which a manned lunar landing took place. A very large rocket system, some 120 m tall, sent a craft containing three astronauts towards the Moon. Small corrections to its path made the spacecraft go into orbit around the Moon, and two of the astronauts then moved into a small, detachable Lunar Excursion Module. Controlled by its own rockets, this small craft moved down and landed on the Moon's surface.

Fig. 26.7 shows the second astronaut, Buzz Aldrin, climbing down from the living space to the surface, photographed by the first astronaut, Neil Armstrong. Meanwhile, the third astronaut, Michael Collins, remained in orbit round the Moon in the Command Module.

After some 21 hours on the Moon's surface, the two astronauts in the Lunar Module lifted off and returned to the Command Module. Fig. 26.8 is a photograph taken by Collins as the Lunar Module approached him. You can see the Lunar Module, the Moon's surface, and a distant view of the Earth. The three astronauts once more in the Command Module, used rocket thrust again to accelerate their craft back towards Earth. In all, there were six Apollo missions which each landed two astronauts on the Moon's surface, to explore different regions. About 380 kg of rocks and dust were brought back to Earth so that detailed analysis could be carried out.

Figure 26.8 The Lunar Module containing the astronauts returns from the Moon's surface to the Command Module, with the Earth visible in the background
(NASA)

26.5 Orbits in space

Metal washers on the end of a string can be whirled round in a circular orbit as in Fig. 26.9. The force needed to keep the mass going round is provided by the tension in the string. In the same way, the Moon is kept going in its orbit round the Earth by the invisible gravitational attraction of the Earth.

Any artificial satellite orbits the Earth similarly, because of the gravitational attraction between the satellite and the Earth. The time a satellite takes to complete an orbit depends on its distance from the centre of the Earth. **Orbits of larger radius take longer to complete**. So, while the first artificial satellites orbited the Earth in about **90 minutes** at a height of a few hundred kilometres above the Earth's surface, the Moon makes a **larger orbit with a period of about a month** at a distance of about 380 000 kilometres.

The same kind of force of gravitational attraction between the Earth and the Sun enables the Earth to orbit the Sun in about 365 days. In the orbits of the Moon and the Earth, the smaller mass appears to move around the larger mass, while each attracts the other.

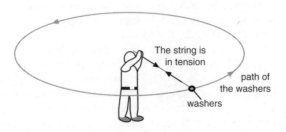

Figure 26.9 Washers are pulled towards the centre of their circular orbit as they are whirled round on the end of a string

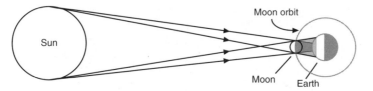

Figure 26.10 A solar eclipse occurs when the Moon's shadow falls on the Earth's surface. In the centre of the shadow the eclipse is total, and at the edge of the shadow the eclipse is partial (Not to scale)

26.6 Eclipses of the Sun and Moon

When the Moon in its orbit passes between the Sun and the Earth, the shadow of the Moon may fall on the Earth's surface. Places in the shadow experience a **solar eclipse** (see Fig. 26.10). Notice that this can only occur when there is a **new moon**.

During a total eclipse of the Sun, the dazzlingly bright disc of the Sun is invisible and a much fainter surrounding glow, called the **solar corona**, can be seen. Sometimes huge sheets of flame, **solar prominences**, are also visible (see Fig. 26.11). You should never look directly at the Sun either with the naked eye or through a telescope. The Sun's bright image formed on the eye's retina can cause permanent injury.

While direct viewing of the Sun during an eclipse is to be avoided, there are ways of recording the view photographically. Astronomers are interested in studying the light from distant stars after it has passed close by the edge of the Sun, and this can best be done during a total eclipse.

Just as the Sun can cast a shadow of the Moon onto the Earth, it can also cast a shadow behind the Earth. When the Moon passes into this shadow, (at a time of **full moon**), there is an eclipse of the Moon or **lunar eclipse** (see Fig. 26.12). From being a brightly-lit disc, the Moon's surface than appears only faintly-lit with a reddish glow.

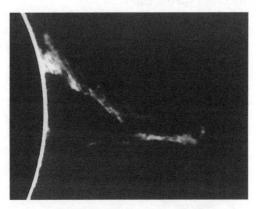

Figure 26.11 A photograph of the Sun's corona during a total eclipse. Prominences hundreds of thousands of miles long can be seen
(Mount Wilson and Palomar Observatories)

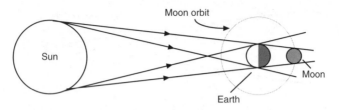

Figure 26.12 *The positions of the Sun, Earth and Moon are shown during a total eclipse of the Moon (Not to scale)*

26.7 Tides

Just as the Earth exerts a gravitational attraction on the Moon, so the Moon exerts a gravitational attraction on the Earth. This force from the Moon is most obvious in its effect on the Earth's oceans; it causes the tides.

The gravitational attraction between objects depends on their distance apart. The greater their separation, the smaller is the force between them. The result of this is that the water on the side of the Earth nearest to the Moon bulges towards the Moon; and the water on the opposite side of the Earth, furthest from the Moon, bulges away from the Moon as in Fig. 26.13.

As the solid Earth rotates, there are therefore two high tides and two low tides each day. The detail of what happens at any place is complex, depending on things like the latitude of the place and the shape of the local shore line. So, while the variation in water depth in deep oceans is about 1 m, it can be more than 10 m in the Bay of Fundy and the Severn estuary. There are virtually no tides in the Mediterranean Sea, as the narrow Straits of Gibraltar prevent much water from flowing into or out from the Mediterranean Sea to change its level.

The Sun also effects the tides on Earth. Although much more massive than the Moon, the Sun is much further away, and its tide-raising effect is only about one third that of the Moon.

When the Sun and Moon are nearly in line with the Earth, at full or new moon, their tidal effects add up. Then the high tides are especially high, and the low tides are especially low. These are called **spring tides**.

In between, when the Moon is at first or third quarter, the tidal effect of the Moon is reduced by the Sun to the maximum extent. The result is that high tides are not so high as the average, and low tides are not so low. These tides are called **neap tides**.

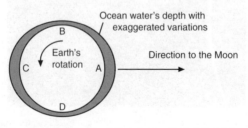

Figure 26.13 *The force of gravity from the Moon causes high tides at A and C, and low tides at B and D. There are two high tides and two low tides every day*

26.8 The planets

There are nine known planets, including the Earth, which orbit the Sun. Many of these planets have moons; Jupiter has at least 17. Each moon revolves around its parent planet under the force of gravity attracting it to the planet. Each planet with its moons, if any, goes round the Sun under the force of gravity attracting it to the Sun.

> **Saturn,** the sixth planet in distance from the Sun, is unique in the solar system. When Galileo first looked at it through a telescope in the 17th century, he saw what was clearly not a simple spherical object. We now know that the planet is roughly spherical, although bulging at its equator; but it is surrounded by a clearly visible ring system made up of billions of small particles, each orbiting the planet in its equatorial plane. Fig. 26.14 is a photograph of Saturn taken with a 150 cm telescope in California, USA.

The various planets are compared in Table 26.1. Notice how the time for a planet to orbit the Sun increases as the average distance from the Sun increases. See also how the surface temperatures and atmospheres on other planets differ from those on Earth.

Only Mercury and Mars have surfaces which can be seen directly through telescopes on Earth. All the other planets are surrounded by clouds which obscure any underlying features. More details have, however, been obtained from data obtained by space probes sent close to the other planets.

Figure 26.14 A system of rings surrounds the planet Saturn
(Palomar Observatory/Caltech)

Table 26.1 A comparison between planets in the solar system

Planet's name	Diameter as multiple of Earth's diameter	Distance from Sun as multiple of Earth/Sun distance	Time for orbit in Earth years	Time for one rotation in Earth days	Atmosphere (main components)	Mean surface temperature in degrees C
Mercury	0.39	0.39	0.24	59	None	−173 to +430
Venus	0.95	0.72	0.62	243	Carbon dioxide	472
Earth	1	1	1	1	Nitrogen and oxygen chiefly	−50 to +50
Mars	0.53	1.52	1.88	1.03	Carbon dioxide chiefly	−140 to +20
Jupiter	11.2	5.2	11.9	0.41	Hydrogen, methane and ammonia chiefly	Cloud top −110
Saturn	9.4	9.5	29.5	0.43	Hydrogen and methane chiefly	Cloud top −180
Uranus	4	19.2	84	0.72	Hydrogen chiefly plus some helium	Cloud top −220
Neptune	3.9	30	165	0.67	Hydrogen and methane chiefly	Cloud top −216
Pluto	0.18	39	248	6.4	Some methane	−230

Figure 26.15 Sojourner explores the surface of Mars
(NASA/Photo Science Library)

Sojourner on Mars

The photograph taken in July, 1997 (Fig. 26.15) shows the robot vehicle 'Sojourner' on the surface of Mars. The vehicle is 63 cm long and 48 cm wide and has a mass of about 9 kg. It was powered by solar panels which allowed a few hours of movement each day.

In the photograph, Sojourner has run down the ramp from its launching platform and is moving towards a rock which it will explore.

The vehicle carried an alpha particle source (see chapter 28), with which to bombard any selected Martian material. From the particles produced by the bombardment, the chemical composition of the material could be deduced.

26.9 Comets

Every so often an object appears in the sky which is quite different from any planet. It is made up of a loose assembly of small particles together with a thin gas, and moves in an orbit round the Sun well outside the Earth's atmosphere. It is a **comet**.

Comets shine mainly by reflecting sunlight. A comet may have a tail made up of material from the main part, the head. As a comet approaches the Sun, the tail develops; then when the comet goes further away again, the tail shortens and finally disappears. The tail is normally directed away from the Sun, driven by the pressure of the Sun's radiation, as in Fig. 26.16.

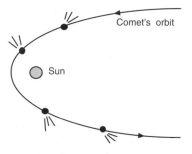

Figure 26.16 The tail of a comet always points away from the Sun. This is caused by the pressure of the Sun's radiation

In the spring of 1997 the bright comet **Hale-Bopp** was easily visible. It had more than one tail. Apart from the main tail, further tails can be formed by ionized gas or atoms driven from the comet's head by charged particles from the Sun. Tails formed in this way are usually narrow and very straight. Hale-Bopp showed a tail due to sodium ions as well as the normal tail containing dust.

The orbits of planets around the Sun are approximately circular. Most comets however have long, narrow, nearly elliptical orbits, so that their distances from the Sun change a great deal during their orbits. **Halley's** comet for example is probably the best known. At its nearest approach it is closer to the Sun than the Earth is, but at its greatest range it is beyond the orbit of Neptune. It takes about 76 years to complete each orbit.

Some comets do not have a repeating periodic orbit. They come near the Sun, swing round it, and go off into inter-stellar space. These comets will never return to the solar system unless their paths are reshaped by some distant interaction.

A comet's orbit is principally determined by the gravity of the Sun; but the orbit can be modified if the comet passes close to one of the planets. The **Shoemaker-Levy** comet, for example, collided with the planet Jupiter in July, 1994, showing us how craters on the Moon's surface may also have arisen. The largest fragments of the comet yielded huge amounts of kinetic energy and created effects like big explosions on Jupiter's surface. The four largest fragments produced plumes of debris which rose about 3000 kilometres above the tops of Jupiter's clouds. No collision of this kind had ever been seen before, and the event gave astronomers a good deal of data about Jupiter's atmosphere.

The major planets, Jupiter and Saturn, are the ones most likely to cause some change in a comet's orbit, as they are by far the most massive planets in the solar system.

Questions

1 What source of light enables us to see the Moon and the planets?
2 Why do we on Earth experience days lasting 24 hours?
3 Why do the lengths of daylight and darkness in Vancouver, say, change during each year?
4 What is the season in Australia when it is mid summer in Canada?
5 At what times of the year will the Sun set 12 hours after it rises?
6 Are there spring tides or neap tides at times of an eclipse (a) of the Sun, and (b) of the Moon?

7 Why can dish aerials on Earth, pointing in fixed directions, receive signals from a satellite?
8 Why do some planets orbit the Sun in less time than one year on Earth?
9 Why are some planets hotter on average than the Earth, while some are cooler?
10 Give one reason why the Moon is not surrounded by an atmosphere while the Earth is.
11 Describe a solar eclipse with the aid of a diagram.
12 (a) Write down one similarity between a comet and a planet.
   (b) Write down one difference between a comet and a planet.

What have you learnt?

1 Can you name the source of light in the solar system?
2 Do you know why the planets are visible to us on Earth?
3 Can you explain the reason we have day and night?
4 Can you explain the sequence of seasons at a point on the Earth's surface?
5 Do you know that only one side of the Moon can ever be seen from Earth?
6 Do you know what keeps some objects in orbit around others?
7 Can you describe a solar eclipse and a lunar eclipse?
8 Can you describe how the Moon and the Sun affect tides on Earth?
9 Can you explain why some planets take longer than others to orbit the Sun?
10 Do you understand how the Moon's appearance changes during a lunar month?
11 Can you describe a comet?

27 The Sun, the stars and the universe

Objectives

After this chapter you should
- know that the Sun appears close to twelve different constellations during a year
- know what could be the Sun's source of energy
- know that the luminosities and surface temperatures of stars can be estimated
- know what is plotted on a Hertzsprung-Russell diagram
- understand that a star can develop through several stages during its long existence
- know why the universe is thought to be expanding
- understand what is meant by the 'big bang' and the 'big crunch'.

27.1 The scale of the universe

In Section 26.1 we said that the distance from the Earth to the Sun (our nearest star) is about 150 million kilometres. The next nearest star is about 300 000 times further away, a distance of about 45 million million kilometres.

Distances to the stars are so enormous that a different scale is often used. Light travels in space at a speed of 300 million metres per second. It therefore takes about 8 minutes to travel from the Sun to the Earth; and at this speed it travels about 9.5×10^{12} kilometres in one year. This distance is called **one light year**. Using this unit we can say the distance to the next nearest star is about 4 light years, as its light takes about 4 years to reach us.

With your unaided eyes you may see about 6000 stars; but with a pair of binoculars you may see about 100 000. The stars are not distributed uniformly. There are many more lying along a faint band of light called the Milky Way, than in directions at right angles to this.

Astronomers have concluded that all these stars and many more (about one hundred thousand million of them, i.e. 10^{11}) together with our Sun, lie in or close to a flattened disc-shaped space called our **Galaxy**. Our Sun is about two thirds of the way out from the centre to the edge of the disc (Fig. 27.1). The diameter of the galaxy is about 100 000 light years, its average thickness is about 4000 light years, and its central nucleus is obscured because large amounts of dust block out its star light. At much greater distances from us, many millions of other galaxies are known to exist. See, for example, Fig. 26.1.

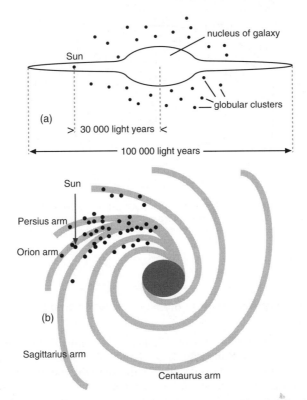

Figure 27.1 The Milky Way occupies a disc-shaped space. Dots outside the disc represent globular star clusters. (b) Radio astronomers have shown that in plan, the stars are arranged in a roughly spiral pattern

27.2 The constellations

When you look at the sky on a clear cloudless night, you can see thousands of stars, some lighter than others, arranged in random groups. These groups are called **constellations**, some of which have been recognised and named for thousands of years. Today astronomers recognise 88 named constellations.

The stars in a particular constellation may only appear to be associated; they may

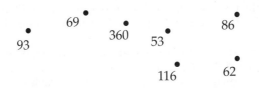

Figure 27.2 The stars in Ursa Major (the Plough) are not close to one another, though their angular separation is small when viewed from Earth. The numbers show their distances from Earth in light years

actually be at very different distances from Earth. For example, Fig. 27.2 shows the distances from Earth of the principal stars in the constellation **Ursa Major** (also called the Big Dipper, the Great Bear or the Plough).

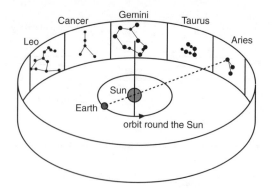

Figure 27.3 As the Earth orbits the Sun during a year, different constellations of stars are in the background. Here the first five of the twelve signs of the Zodiac are indicated, as if the constellations were all on a cylindrical surface at the same distance from the Sun

---

Self test

Why do these stars appear to stay in the same formation seen from Earth?

As the Earth moves in orbit round the Sun, a succession of different constellations form the background to the Sun (Fig. 27.3). None of these constellations can be seen at the time, because any observer is dazzled by the sunlight coming from that direction. But knowing the positions of the stars, seen for example six months earlier, their positions in the background to the Sun are known. The stars through which the Sun appears to move in one year are divided into twelve constellations, the so-called '**signs of the zodiac**'.

Self test

Why does each sign of the zodiac relate to a particular time of year?

27.3 The Sun's source of energy

It has been calculated that if the Sun had been made of the best coal, set alight 5000 years ago, and supplied with oxygen to keep it burning, it would all have burned up by now. But it shows no sign of burning out; the Sun's energy does not come from chemical reactions.

The Sun's energy derives from **nuclear reactions. Hydrogen** and **helium** are the main constituents of the Sun and the stars. At the very high temperatures deep inside stars, pairs of hydrogen nuclei can fuse together, each pair forming one helium nucleus and at the same time releasing energy (see Section 29.4) This **fusion reaction** between nuclei is also called a **thermonuclear reaction**. If it were possible to control the rate of this reaction on Earth, the hydrogen in the oceans might one day become a cheap fuel for a new type of nuclear power station (see Section 28.4).

27.4 Cataloguing stars

Surface temperature. If you are sharp-eyed, you may notice that stars are not all the same colour. While they all look approximately like white points of light, some are redder and some more blue than the average. The differences can be demonstrated using astronomical instruments.

To explain the differences, we can use a single-bar electric fire which has a shiny metal reflector. After the fire is switched on, you can feel some heat coming from it before there is any visible change; the fire is then emitting infra-red radiation. As it gets hotter, the electric element first becomes dull red, and then a more orange-red. Objects which become even hotter than the fire can become white-hot.

The colour appearance of an object in space is a guide to its temperature, and the analysis of the radiation coming from a star, including the quantities of radiant energy of different wavelengths, enables the star's surface temperature to be determined.

Luminosity. Stars seen from Earth differ in brightness; some are brighter than others. If one star (A) appears fainter than another star (B) it could be that **A is further away than B**, or that, although A and B are equally far away, **A is giving out less radiation than B**, perhaps because **A is smaller than B**.

To compare the brightness of stars, astronomers work out how bright they would appear if they were all at a standard distance from the Earth. They also make a correction to allow for the invisible radiation emitted by the star as well. This produces the so-called **absolute magnitude** of each star if we could see all its emitted radiation. Then they convert the absolute magnitude of each star into its **luminosity**. The Sun's luminosity is defined as 1, and a star with a luminosity of 6.0 would appear 6 times as bright as the Sun when the same distance away.

Stars for which the surface temperature and luminosity are known can be plotted on a **Hertzsprung–Russell diagram** (H–R diagram), named after the astronomers who first used it (Fig. 27.4).

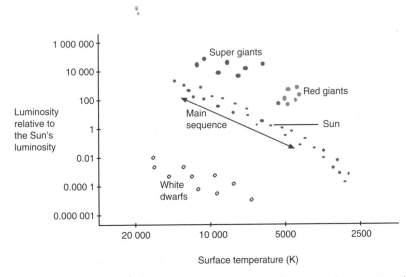

Figure 27.4 The general appearance of the H–R diagram includes the Main sequence, which contains about 90% of all stars. The others in distinct regions have special characteristics

27.5 The H–R diagram

About 90% of all the stars on the diagram lie in the region labelled **Main sequence**, running from the top left to the bottom right. Our Sun is one of these stars. As you might expect, hotter Main sequence stars are more luminous than colder Main sequence stars.

Other stars are very different. Some lie at the top right of the diagram. These are more luminous than Main sequence stars of the same temperature, because they are much larger. One of these identifiable **giant stars** is Betelgeuse in the constellation Orion (the top left corner star, above the belt and sword in Fig. 27.5). Betelgeuse is so

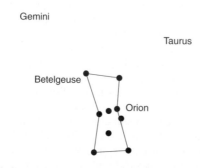

Figure 27.5   Betelgeuse is the star at one 'corner' of the constellation Orion. Taurus and Gemini are two more constellations next to Orion

big that if it were to replace our Sun, the planets Mercury, Venus and Mars, as well as the Earth, would all be engulfed by it!

In the bottom left of the H–R diagram are stars of another quite different type, being very faint and very hot. They are small stars called **white dwarfs**.

27.6 The life of a star

The life span of a star is thought to be many billions of years, so no one can be certain of its evolution. Nonetheless astronomers have managed to suggest a possible sequence, based on the laws of physics which apply on Earth.

A star is thought to form from a spread-out cloud of gas and dust, which contracts by gravitational attraction. As it shrinks losing gravitational PE, the mass heats up (as gas in a bicycle pump heats up when compressed). Eventually its interior gets up to a critical temperature of about 1 million degrees kelvin, at which the fusion of hydrogen nuclei into helium nuclei, described in Section 27.3, can begin. The star then settles down on the Main sequence.

The length of time spent on the Main sequence may be billions of years. When the amount of hydrogen in the star becomes low, further changes in the star begin. The central core of the star shrinks, thereby raising its temperature even more, so that different nuclear reactions in the core become possible and more energy is released. The result is that the outer layers of the star expand, so that the surface temperature drops. However, the star's luminosity becomes greater because of its increased surface area. The star has now become a **red giant**, and has evolved away from the Main sequence.

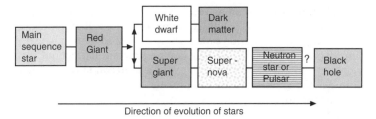

Direction of evolution of stars

Figure 27.6 *Possible stages in the life of a star, from left to right*

Beyond this point there are two distinct paths of evolution. If the star is about as massive as the Sun or less, it reaches the point where it has insufficient hydrogen for the fusion process to continue. It is no longer supported by the continual generation of energy in the core. It collapses fairly rapidly, thereby losing some gravitational potential energy which is released as radiation. This results in a star which is small and densely packed, quite hot but gradually cooling as its energy supply is diminishing; it is a **white dwarf** (see Fig. 27.6). Finally it cools sufficiently to become invisible; it becomes part of the dark matter which astronomers suspect exists in space even if they cannot see it.

If the original star is much more massive than the Sun, its evolution may be different. At the **giant or super giant** stage, such a star becomes unstable, expanding and contracting several times as further fusion reactions lead to the formation of nuclei of heavier elements. The star then becomes a **supernova**, **blowing itself apart** with the release of a huge amount of energy. It also throws its outer layers of dust and gas into the surrounding space.

Supernovae are rarely seen. In 1054 AD Chinese astronomers recorded seeing a bright star during daylight where, at night, we now see the Crab Nebula. Today the remains of this supernova appear as a faint cloud of expanding hot gas and debris shown in Fig. 27.7. Two supernovae, one reported by Kepler, were observed in the

Figure 27.7 The Crab Nebula, the remains of a supernova explosion, is a cloud of expanding hot gas. The central remnant of the original star is thought to be a neutron star
(Palomar Observatory/Caltech)

16th century; and another was seen as recently as 1987 in a nearby star cluster. The brightest phase of a supernova lasts only a matter of days, and supernovae occur in a galaxy on average about once in 500 years.

After a supernova explosion occurs, the central remnant left behind is very densely packed, more so than in a white dwarf. It is called a **neutron star** or **pulsar**; 'neutron star' because under the great gravitational pressure in the star, protons and electrons get squashed together to form neutrons; 'pulsar' because these stars rotate rapidly while emitting radio waves, so that pulses of radio waves from them can be detected on Earth.

If the mass of the neutron star is large enough, the star may be crushed by its own gravity into a smaller and smaller volume. Eventually as its volume becomes very small, the gravitational field strength at its surface becomes so great that nothing can escape from the star, no object and no radiation. The star has become totally invisible; it is then called a **black hole**. Fig. 27.6 illustrates a possible sequence of stages in the life of a star. There is however, a suggestion that some massive stars may collapse to black holes without producing supernova-type explosions.

27.7 The expansion of the universe

You will be familiar with the effect caused when a car with a siren moves towards you. As the car approaching you speeds up, the pitch of the note you hear goes up, and when it moves away from you, the pitch of the note goes down. This effect in sound is called the **Doppler Effect**.

Exactly the same kind of effect occurs with light. If a light source is moving towards you, the light 'pitch' goes up, and the light appears more blue (blue light has a higher frequency than red light). If the light source moves away from you, its light appears more red, i.e. the frequency is lowered.

Early in the twentieth century astronomers noticed a Doppler Effect in light reaching us from distant galaxies. The light from stars in galaxies a long way off appears shifted towards the red end of the spectrum; and the further away the star is, the more its light displays a shift towards the red. This is interpreted as meaning that distant galaxies are receding from us, and the further away they are, the faster they are going. Edwin Hubble published the results of his study of this effect in the form of a summary known as **Hubble's law**. The law states that **the velocity of recession of a galaxy is proportional to its distance from us** (see Fig. 27.8).

It seems therefore that the distant galaxies are moving away from us and from each other. The universe is expanding. It follows that some billions of years in the past the galaxies must have been much closer together. Many astronomers think that the whole

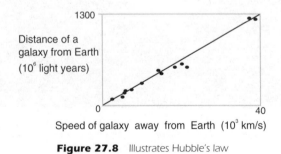

Figure 27.8 Illustrates Hubble's law

process of expansion which we can observe began with a massive explosion, called the **Big Bang**, when all the material in the Universe began to expand from its original very crowded, very dense and very hot state.

If we know how far distant galaxies are from us, and if we know their speed away from us, it is possible to calculate how long ago the Big Bang took place. In other words, we can calculate the age of the Universe; it has been calculated as about 10^{10} Earth years.

Self test

What assumptions are being made in this last calculation?

27.8 What of the future?

There is at the moment some uncertainty about how much material the whole universe contains. As the total amount would alter theories about future developments in the universe, astronomers can only speculate about its future evolution.

Astronomers think there is a great deal of matter we cannot see although it is there. Some suggest that we can only see about 10% of what is actually present; the rest is referred to as **dark matter**, which is invisible as it gives out no radiation. If the whole universe contained enough material, including the dark matter, then the expansion of the galaxies away from each other would slow down under mutual gravitational attraction. Eventually expansion would cease and the whole process would go into reverse. Ultimately all material would contract into a smaller and smaller space, ending in what has been called the **Big Crunch.**

If there is not enough material for gravitational attraction to stop the expansion, then the Universe will continue to expand; and its galaxies will go on moving apart indefinitely.

Questions

1 What is a light year?
2 What happens in a thermonuclear reaction?
3 What must be studied in order to estimate a star's surface temperature?
4 Give two reasons why one star may appear to be brighter than another.
5 What happens first when a star's nuclear fuel begins to run out?
6 (a) What type of energy is lost when the outer mass of a star collapses inwards?
  (b) How can this lead to the star becoming more luminous than it was?
7 What type of star will eventually become a white dwarf?
8 What do we mean by a supernova?
9 (a) What particles must be squashed together to form neutrons?
  (b) How could this happen to form a neutron star?
  (c) Why are neutron stars sometimes called pulsars?
10 What do we mean by a black hole?

Nuclear physics

28.1 Introduction

At the heart of a nuclear reactor, energy from the nuclei of atoms is transformed into heat. The heat can be used in a power station to generate electricity, or in a ship's propulsion unit; and there are many further applications of nuclear energy on a smaller scale. The properties of atomic nuclei which can give up this energy are discussed in this chapter.

28.2 The nuclear atom

An important experiment was performed by Geiger and Marsden in 1911. They projected a beam of fast moving particles, called alpha (α) particles, at a very thin piece of gold foil.

It was expected that the α particles would pass straight through the foil, and many did. But some were deflected through quite large angles, and a few bounced back as in Fig. 28.1 onto the same side as the source of alpha particles.

Rutherford produced a model to explain these results. He suggested that an atom consists of a very small positively charged nucleus in which almost all the mass of the atom is concentrated. The nucleus causes α particles to be deflected (see Fig. 28.2). Spread out in the space surrounding the nucleus are the electrons, which carry

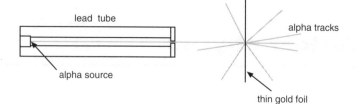

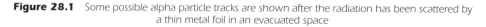

Figure 28.1 Some possible alpha particle tracks are shown after the radiation has been scattered by a thin metal foil in an evacuated space

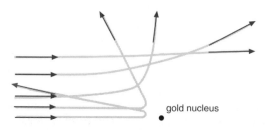

Figure 28.2 A few alpha particle tracks are deflected through large angles where they approach a gold nucleus. The tracks furthest from the nucleus show the least deflection

negative charge. They do not deflect α particles, as an electron has a much smaller mass than an α particle.

Geiger and Marsdens' results agreed closely with Rutherford's predictions, and his model of the nuclear atom is the one we use today.

28.3 The make-up of the nucleus

In 1920 Rutherford suggested that two types of particle would be found in the nucleus of an atom. These were called the proton and the neutron. In 1932, Chadwick found experimental evidence for the neutron. It has the proton's mass, but the neutron is completely uncharged, while the proton has a positive charge equal and opposite to the negative charge of the electron.

The number of electrons in an uncharged atom is matched by the number of protons inside its nucleus. As an example consider a neutral atom of carbon. It has six electrons in the space around its nucleus. That means that it has six protons inside its nucleus; but it also has six neutrons which add to its mass. These are listed in Fig. 28.3.

Two whole numbers describe the nucleus of an atom. The first is the **atomic number** or proton number. The atomic number of carbon is six because all its atoms have six protons in their nuclei, balancing the charge of the six surrounding electrons. This is also why carbon is number six in the chemical periodic table.

The second number used to describe the nucleus is its **mass number**. This is the **total number of protons and neutrons** which make up the nucleus. The carbon atom described above has six protons and six neutrons, so its mass number is 12. The chemical symbol for carbon is C, and the nucleus of this atom is described by the

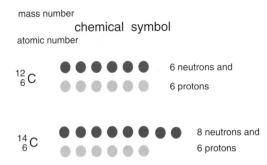

$^{12}_{6}$C   6 neutrons and 6 protons

$^{14}_{6}$C   8 neutrons and 6 protons

Figure 28.3 In each isotope of carbon the atomic number (6) is the number of protons in the nucleus. The mass number is the total number of protons and neutrons in each nucleus

numbers to the left of the symbol. So $^{12}_{6}$C describes the nucleus. Twelve is the mass number, and six the atomic number. All nuclei can be described with a mass number and atomic number. $^{4}_{2}$He for example is a nucleus with 2 protons and 2 neutrons, whereas $^{235}_{92}$U has 92 protons and 143 neutrons.

28.4 Isotopes

Some nuclei of carbon contain eight neutrons instead of six. They behave in exactly the same way in chemical reactions, but their nuclei have the extra mass of the two extra neutrons.

$^{14}_{6}$C is the description of the more massive 'carbon fourteen' nucleus. A very few carbon atoms have seven neutrons, and so a mass number of thirteen. The three types of carbon atom are called **isotopes** of carbon. Other elements also have isotopes. The isotopes of any element have the **same atomic number, but different mass numbers**, because they have the same number of protons but different numbers of neutrons in their nuclei.

28.5 Some nuclei emit radiation

Most naturally occurring atomic nuclei are stable, but some emit radiation. This is called **nuclear decay**, and it leaves the original nucleus changed in some way. If the new form of the nucleus is stable, it does not emit any more radiation; but if the new form is unstable it may emit radiation again after some time.

The three main types of radiation from nuclei are called alpha, beta and gamma radiation, sometimes using the Greek letters α, β and γ. They have some common properties:

(a) they all possess energy and can penetrate matter to some degree
(b) they can be detected in the dark by means of a photographic plate
(c) they can ionise neutral atoms.

They can be distinguished using the apparatus shown in Fig. 28.4. The least penetrating is called α radiation, and the most penetrating is γ radiation.

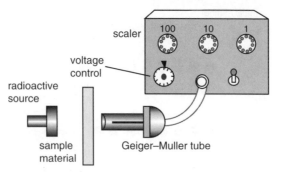

Figure 28.4 The penetrating powers of nuclear radiations in various materials is found using a G–M tube and scaler

28.6 Detecting radioactivity

Most detectors of nuclear radiation make use of its ionising property. The best known detector is the **Geiger–Muller tube** (G–M tube), named after its originators and shown in Fig. 28.5. The tube has a fine wire running along the axis of a metal cylinder. The cylinder, which contains a vapour at low pressure, has a thin window at one end to allow less penetrating radiations like α radiation to enter. An electrical power supply

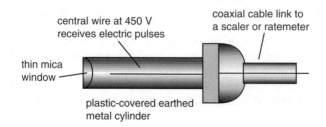

Figure 28.5 A Geiger–Muller tube has a low pressure vapour in a cylinder containing a central wire at positive voltage. The thin mica window allows radiation to penetrate the tube easily

of perhaps 450 volts is connected between the cylinder and the wire, so that ions created by α, β or γ radiations are attracted to one part of the tube or the other. Then, for a short time a small current flows, and the pulse of current is counted automatically by an electronic pulse counter or scaler. A pulse is counted each time an ionising radiation passes through the tube. Alternatively the G–M tube could be connected to a ratemeter which shows the number of electric pulses counted per second.

28.7 Radioactive emissions

An unstable nucleus can break apart spontaneously, and the resulting nucleus has less energy and is usually more stable than its parent nucleus.

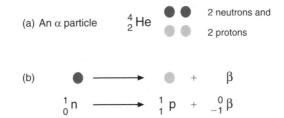

(a) An α particle $^{4}_{2}$He   2 neutrons and 2 protons

(b) ● ⟶ ○ + β

$^{1}_{0}$n ⟶ $^{1}_{1}$p + $^{0}_{-1}$β

Figure 28.6 *Figure (a) shows the make-up of an alpha particle. (b) A neutron inside the nucleus can change into a proton and an electron which is emitted as a beta particle*

(a) α *particle emission*

An α particle consists of two protons and two neutrons bound as a single particle. It is the nucleus of helium, and may be written: $^{4}_{2}$He.

When it is emitted from a nucleus, the number of protons and neutrons in that nucleus both decrease by two (see Fig. 28.6a). So the atomic number falls by two and the mass number by four. An example of the change resulting from α particle emission is:

$$^{238}_{92}U \longrightarrow {}^{234}_{90}Th + {}^{4}_{2}He$$

Notice that 238 = 234 + 4. Also see that 92 = 90 + 2. So the atomic numbers before and after emission are balanced, and so are the mass numbers. Here, a nucleus of uranium has changed into one of thorium, by the emission of an α particle.

(b) β *particle emission*

β radiation consists of electrons. In an unstable nucleus which emits β radiation, a neutron changes into a proton and an electron. Of these, the electron is emitted from the nucleus and the proton remains (see Fig. 28.6b). A new nucleus is formed with one more proton and one fewer neutron. The atomic number increases by one but the mass number stays the same. e.g.

$$^{24}_{11}Na \longrightarrow {}^{24}_{12}Mg + {}^{0}_{-1}e$$

Here the unstable sodium nucleus changes into an isotope of magnesium, and an electron is emitted from the nucleus as a β particle. The mass of the electron is so small compared with the mass of a proton that it is given a mass number of 0.

(c) γ *emission*

γ radiation has neither charge nor mass, so it does not cause any change in the charge or mass of the emitting nucleus. It does, however, carry away energy in the form of very penetrating waves rather like X-rays, but having shorter wavelengths.

Radiotherapy is a hospital treatment which can use γ-rays to destroy harmful cancer cells in the body. The radiation is so penetrating that thick concrete or lead screening is needed to reduce its intensity where it is not being applied.

28.8 Safety

As we have seen, nuclear radiations can cause ionisation by knocking electrons out of atoms or molecules. They can therefore modify the structure of living cells, causing the cells to die. Workers with radioactive materials need protection to work safely. Here are some of the precautions which can be taken:

(a) radioactive materials are never moved by hand, but remotely, using long tweezers or a machine, which may be observed through a very thick lead-glass window
(b) workers may wear protective clothing including gloves, headgear and boots, which may be hosed down and decontaminated at the end of the day's work
(c) workers carry a sensitive strip pinned to their clothes to show the amount of radiation they have experienced
(d) radioactive sources are stored in steel boxes lined with lead. The lead is an effective absorber of radiation and the steel gives strength to the container. The store room has thick concrete walls
(e) radiation detectors are mounted in different parts of a building where there are radioactive materials. They monitor radiation levels in the air and provide a warning of any leakage.

The sources used for school demonstrations of radioactivity are very weak. Nevertheless, they are stored in lead containers under lock and key, and are handled with long tweezers when in use in the laboratory.

28.9 Further properties of radioactive emissions

α particles produce many ions in air, and so they rapidly lose energy and stop. Their range in air is only about 10 cm, and they can be completely stopped by a few sheets of paper. α emitters are therefore quite safe if they remain outside the human body because the skin acts as a paper-like protection. The only danger arises if they are taken inside the stomach or lungs.

β particles cause less ionisation and have a range in air of a few metres. Although they are more penetrating than α particles, they can be stopped by a few millimetres of a metal like lead or aluminium.

γ radiation produces the least ionisation, and it is more penetrating than β radiation. It readily passes through many metres of air, through skin and bone, and is never totally blocked. Its intensity is reduced to very low levels by thick walls of lead.

Table 28.1 provides a general summary of the properties of nuclear radiations, including their behaviour in regions close to magnets and surfaces carrying electric charges. It is worth noting that γ-rays, which travel at the speed of light, are not affected by electric or magnetic fields. α particles which carry the positive charge of two protons are attracted towards a negatively charged surface, and β particles with negative charge are attracted towards positive charges. Both α and β particles are deflected by magnetic fields much as moving positive and negative charges would be; but α particles are deflected less, as they have relatively larger masses. Fig. 28.7 shows apparatus for demonstrating the magnetic deflections of α and β radiations.

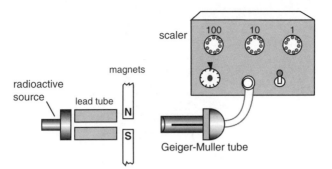

Figure 28.7 *Magnetic deflection of a narrow beam of nuclear radiation can be investigated with this apparatus*

Table 28.1 Properties of α, β and γ radiation

Radiation	Nature	Range in air	Penetration	Electric or magnetic fields
Alpha (α)	Helium nuclei $^4_2$He	Up to about 15 cm	Blocked by a few sheets of paper	Deflected as a moving positive charge would be
Beta (β)	Electrons. They have negative charge $^0_{-1}$e	Up to about two metres	Blocked by a few mms of lead or aluminium	Deflected as a moving negative charge would be
Gamma (γ)	Very short wavelength waves without mass or charge. They travel in space at the speed of light	Much further than either α and β can go	Intensity is much reduced by metres of concrete, or several cms of lead	No effect

28.10 Background radiation

A Geiger–Muller tube and counter registers a small count rate wherever the apparatus is set up. The count rate is due to **background radiation** which is all around us. Over 50% of background radiation comes from naturally occurring atomic nuclei in rocks like granite, or in organic materials. Some of it results from nuclear weapons testing, or from accidents with nuclear reactors as at Chernobyl. Around 10% of background radiation comes from outer space and is called **cosmic radiation**.

In experiments with radioactive materials, allowance has to be made for background radiation, if the effects of the test materials are to be correctly recorded.

28.11 The law of radioactive decay

For any particular radioactive sample the rate of decay follows the same pattern.

A convenient experiment uses radon gas which emits α particles. The gas is puffed into a sealed chamber containing a G–M tube connected to an external ratemeter as in

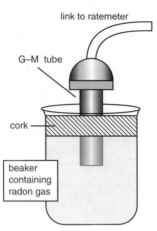

Figure 28.8 *Apparatus used for finding the half life of radon gas*

Fig. 28.8. The ratemeter indicates the rate at which the radon in the chamber is decaying, and Fig. 28.9 shows how the ratemeter reading varies with time.

From the graph it is clear that the **count rate takes the same time to fall to half its original value**, no matter what the original value is. This behaviour, the halving of radioactivity in a fixed period of time, occurs with samples of all radioactive isotopes. The time for halving to happen is called the **half life**, τ, for the process. So τ is the time taken for the activity of a radioactive sample to fall to half its original value.

The values of half lives vary from one isotope to another. They have values ranging from millions of years to fractions of a second. The half life of radon gas in the experiment above is about 56 s. But the importance of half life is that it enables predictions to be made about the activity of a sample in the future, based on its present level of activity.

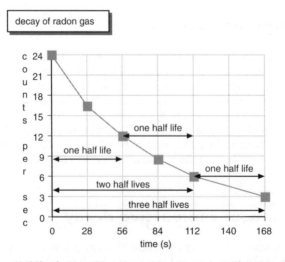

Figure 28.9 In one half life of radon (56 s), the activity reduces by half, whichever starting point you choose

For example, a sample of radioactive cobalt, with atomic number 27 and mass number 60, emits γ radiation with a half life of five years. If initially there are eight million emissions per second, then in five years time the activity of the sample will have halved to four million emissions per second. In a further five years the activity will have halved again to two million emissions per second, and so on. In a time span of three half lives the activity falls to $0.5 \times 0.5 \times 0.5$ of its original value. i.e. one eighth of its starting value.

28.12 The randomness of radioactive decay

Radioactive decay has parallels in many other purely random processes. Imagine you throw a large number of dice quite randomly. Each dice has six faces, so the chance is one in six that any dice will show a particular face upwards. Out of 600 dice we would expect about 100 to show sixes and the other 500 to show other numbers. In a game we could play, a six means a 'decay', and the dice showing sixes play no further part. So after one throw we have about 500 dice still in the game. We say 'about 500' because that is the best estimate we can make. If 'decays' are random, then other results are also possible.

The process is repeated. When the 500 dice are thrown, we expect about 83 to be sixes, so only about 417 of the original 600 dice would remain. The process is continued again and again. Fig. 28.10 shows the results of a typical game plotting the num-

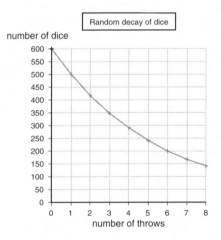

Figure 28.10 *The dice remaining in the game decrease just as the activity of a radioactive sample decays*

ber of dice remaining against the number of throws. If throws are regarded as equal time intervals, the results look very similar to the results for radioactive decay. Yet these results were achieved quite randomly.

On the large scale, the size of the population of dice has reduced in an ordered and predictable fashion, yet the decay of each of the 600 dice was quite random and unpredictable. In the same way individual nuclear decays are unpredictable, but the activity of a radioactive sample, containing many millions of these nuclei, is ordered and predictable. In a certain time we can predict how many nuclei will decay, but we cannot tell which ones will decay.

28.13 Radioactive carbon dating

One of the applications of radioactive decay is in the dating of samples of organic matter laid down some thousands of years ago. The samples could be of wood cut from ancient objects. The isotope, carbon 14, is a β emitter with a half life of about 5700 years. Initially, when the organism is alive, the ratio of its carbon 14 to carbon 12 nuclei would match that of the air in which it grows. On dying, an organism stops growing new cells, and gains no new carbon atoms. In a sample containing carbon, the ratio of carbon 14 nuclei to the stable carbon 12 nuclei gradually reduces after its death, because the number of carbon 14 nuclei decreases, while the number of carbon 12 nuclei stays the same. 5700 years after an organism has died, the amount of carbon 14 it contains will have halved. Measurements using a mass spectrometer to find the ratio of the number of carbon 14 to carbon 12 atoms in a sample therefore enable the time-lapse since its death to be estimated.

28.14 The ages of rocks

The time since a rock solidified from the molten state may be estimated if it contains some radioactive substance. Two examples are given.

Uranium is naturally radioactive. The uranium 238 isotope, ^{238}U, decays via a chain of events to form an isotope of lead, ^{206}Pb; and ^{235}U nuclei decay via another chain to form ^{207}Pb. Now common lead has isotope abundances as follows: ^{204}Pb, 1%; ^{206}Pb, 25%; ^{207}Pb, 22%; and ^{208}Pb, 52%. However, a rock crystal containing uranium has a greater abundance of ^{206}Pb, and ^{207}Pb. The suggestion is that these extra amounts of lead result from the decay of uranium in the sample. We need to know how much of each isotope of uranium is present as well as the amounts of each isotope of lead.

A mass spectrometer is used to determine the proportions of the four lead isotopes present in the sample. From the amount of ^{204}Pb present (not formed by a radioactive decay chain), the amounts expected in common lead with mass numbers 206, 207 and 208 can be calculated. If measurements show that more than these quantities are present, the extra amounts are thought to have been produced by radioactive decay since the rock was formed. So the amounts of ^{206}Pb and ^{207}Pb resulting from the decays of ^{238}U and ^{235}U respectively, can be found.

The half life of the decay from ^{238}U to ^{206}Pb is 4.5×10^9 years; ^{235}U decays to ^{207}Pb with a half life of 7.1×10^8 years. So from the mass spectrometer measurements the apparent 'age' of the rock sample can be calculated.

In a simple example suppose there are equal numbers of atoms of ^{238}U, and ^{206}Pb formed by decay in a sample. Then the time since the rock solidified must be 4.5×10^9 years, the half life for the decay process.

(b) Another process used to 'age' rocks involves potassium nuclei, ^{40}K, which decay into argon nuclei, ^{40}Ar, with a half life of 1.3×10^9 years. The amounts of each isotope present in a sample are determined, and in a manner similar to that described above, it is possible to calculate the age of the rock. For example, if there are found to be equal numbers of ^{40}K and ^{40}Ar atoms present, then half of the original ^{40}K nuclei must have decayed since the rock formed, so its age is 1.3×10^9 years.

1 Name two particles contained in the nucleus of an atom.
2 What type of charge is held in an atomic nucleus?
3 What other particles do atoms contain outside their nuclei?
4 What is the atomic number of an atom?
5 What is the mass number of a nucleus?
6 What word describes two atoms of an element having different mass numbers?
7 What is the half life of a radioactive isotope?
8 Use Fig. 28.10 to estimate the half life of the dice throwing process.
9 What properties are common to α, β and γ radiations ?
10 What are the differences between α, β and γ radiations?
11 What safety procedures are used in the handling and storage of radioactive materials?
12 How many protons and how many neutrons make up one nucleus of:
(a) ^1_1H, (b) ^2_1H, (c) $^{238}_{92}U$, and (d) $^{90}_{38}Sr$?
13 $^{232}_{90}Th$ is an alpha particle emitter. What are the atomic and mass numbers of the resulting nucleus?
14 Why might exposure to nuclear radiations be dangerous?

What have you learnt?

1 Do you understand the structure of an atom in terms of protons, neutrons and electrons?
2 Can you explain what isotopes are?
3 Do you know the characteristics of α, β and γ radiations?
4 Do you know the sources of background radiation?
5 Do you understand the meaning of half life, and can you use it to predict the amount of radioactivity of a source in the future?
6 Do you know the effects of radioactivity on living cells?
7 Do you know how radioactive materials are handled and stored?
8 Can you explain the principles of radio carbon dating?

29 Using radioactivity

29.1 Nuclear fission

We stated in Section 28.14 that naturally occurring uranium is radioactive; it emits alpha particles. But one uranium isotope can also decay by a different process called **spontaneous fission**. The nucleus can break up spontaneously into two large fragments and a few neutrons, with the release of energy, in the form of kinetic energy, of all of the pieces. The isotope which can do this is uranium 236, or ^{236}U.

Almost all uranium is made up of two isotopes, each with atomic number 92, but with mass numbers of 238 (over 99%), and 235 (0.7%). Spontaneous fission is a rare event, but it is more likely if slow-moving neutrons are captured by nuclei of uranium 235 atoms. Neutron capture by a nucleus of uranium 235 produces uranium 236, which is unstable and quickly undergoes fission. More energy is released as more neutrons are captured by uranium 235 nuclei, producing more fission, and so on. This is called a **chain reaction**.

Unfortunately faster neutrons can be captured by uranium 238 nuclei without causing any fission. We need to slow down the neutrons in a nuclear reactor, or very few ^{236}U nuclei will be created, and very little spontaneous fission will occur.

29.2 Nuclear reactors

The essential fuel is uranium, which is sometimes 'enriched' by increasing the proportion of the uranium 235 isotope. The fuel is often produced in the form of rods contained in sealed metal cans.

When a quantity of enriched uranium is assembled, spontaneous fission takes

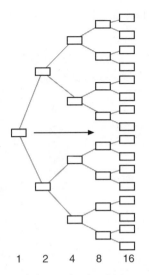

1 2 4 8 16

Figure 29.1 This figure illustrates a chain reaction in which, moving from left to right by stages, the number of small boxes (or dominoes) doubles from one stage to the next

place somewhere, in one of the nuclei. This releases some neutrons, which begin the chain reaction described above. As the chain reaction builds up and more neutrons are being captured, energy is released more rapidly. In Fig. 29.1 dominoes representing neutrons increase in number at each stage moving to the right.

The rate of energy release in a nuclear reactor has to be controlled, so that an explosion or melt-down cannot occur. The control is achieved by:

1 keeping the assembled mass of uranium below a certain limit
2 spacing the fuel elements so that there are not too many lone neutrons in the reactor
3 absorbing unwanted neutrons in the spaces between fuel rods.

Boron rods absorb neutrons without becoming radioactive. Lowering the rods into the reactor reduces the rate of the chain reaction; and raising the rods increases it.

The moderator in the reactor **slows down the neutrons.** Without it becoming radioactive, the moderator's presence means that more neutrons are likely to be captured and cause fission. Carbon is often used as a moderator (see Fig. 29.2).

The purpose of a reactor is to transform nuclear energy into heat, which can be used for generating electricity, for example, in a power station. As fast as the heat energy in the reactor is produced, it is carried away by a heat exchanger containing a fluid circulating through the reactor in a closed loop of pipes. The heat exchanger delivers heat to a boiler where water is boiled. Then steam at high pressure drives the turbines which operate the generators. The stages of the process are:

1	→	2	→	3	→	4	→	5
Nuclear energy from fission causes heating		A heat exchanger transfers the heat energy to a boiler		Hot water boils to become high pressure steam		High pressure steam drives the turbines		Turbines drive electricity generators

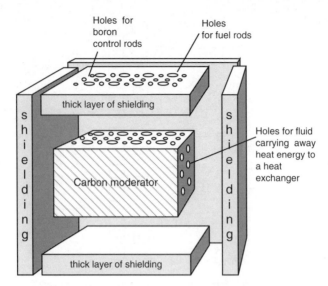

Figure 29.2 A nuclear reactor has channels for fuel rods, control rods and for a fluid involved in heat transfer

29.3 The nuclear debate

Attitudes towards the use of nuclear energy in power stations vary from one person to another. Governments ultimately decide on the future of nuclear energy programs, and so it is not just scientists who need to know some relevant details.

Politicians and Civil Servants must balance a country's energy supplies with its expected needs. For each available resource, several factors must be considered, e.g.:

1   safety factors,
2   costs associated with using each form of energy,
3   the time it will take to use up known reserves,
4   amount of employment it offers,
5   availability of a suitable workforce,
6   the level of public confidence and approval,
7   longer term problems which arise, and
8   other relevant political issues.

Each of these factors can be subdivided, as there are many aspects of safety, or costing, etc.

It could prove a valuable exercise to consider these factors as far as nuclear energy is concerned. A list of advantages and disadvantages of using nuclear energy could be drawn up, so that reasoned conclusions could be reached.

As known fossil fuels are progressively used up, the move towards using more nuclear energy may gain momentum. However, public confidence is affected by the possibility of a leak of radioactive material after an accident such as at Chernobyl in 1986. There is also the possibility of the release of low-level radiation from any nuclear installation.

A long-term problem arising from the use of nuclear energy in a fission reactor is the difficulty in disposing of the spent fuel rods. Although much of their nuclear energy is released in the reactor, there remains active material in the rods, some of which has very long half life. When spent fuel rods are removed they have to be stored, as no one has yet developed a way of disposing of them safely. They could be enclosed in thick, tough casing, but their active material would probably remain active long after the casing had disintegrated. So even storage in disused coal mines presents a risk of contaminating underground water reservoirs.

A further problem arises when nuclear reactors are decommissioned after their useful lives have finished. This can be due to radioactivity in the reactor assembly. Decommissioning adds to the cost of using nuclear power, and leaves a health hazard that is extremely difficult to eliminate.

The increased use of energy from alternative resources could be preferable to an increase in either fossil fuelled or nuclear fuelled plants. The risks of atmospheric pollution and environmental damage could be factors which increasingly shape public attitudes to the choice of power sources.

29.4 Nuclear fusion

Another nuclear process releases energy. It is called **nuclear fusion** because it involves the fusing together of two hydrogen nuclei, typically each with one proton and one neutron, to form a helium nucleus. This process occurs at high temperature, and it results in a large release of energy. It is thought this is the way energy is released in the Sun and the stars (see section 27.3).

The only man-made device using this principle to date is the hydrogen bomb. For nearly fifty years, scientists and engineers have been trying to make fusion occur at a slower, controllable rate, but so far without success.

29.5 Other uses of radioactivity

Apart from radio carbon dating described in chapter 28, there are many applications where small quantities of radioactive material can be used. We shall point to a few applications which indicate a variety of uses.

α radiation

A smoke detector may enclose a small quantity of an α emitter which has a long half life. This is combined with an ion detector which detects the ions created by alpha emission in the air between two charged metal plates. Smoke particles are large enough to reduce the intensity of the radiation; so that, in the presence of smoke, the ion detector gives a smaller output signal. When the signal strength changes significantly, loud 'bleeps' are heard. The safety of this system is that a failing battery also causes the alarm to sound in short 'bleeps', so that the battery can be replaced.

β radiation

A paper mill produces paper whose thickness is controlled by the pressure of rollers (see Fig. 29.3). As the paper emerges from the rollers it passes between an β source and a detector. The signal from the detector is used to control the pressure exerted on

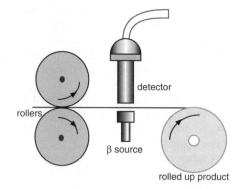

Figure 29.3 A β source is used with a detector to control the pressure of rollers in a paper mill

the paper by the rollers. So, for example, if the paper gets too thick, more β particles are absorbed and fewer reach the detector. The smaller output signal from the detector controls a press which increases the pressure on the rollers, and this in turn reduces the thickness of the paper. The control is continuous and automatic.

γ *radiation*

This is, you will remember, the most penetrating of the three radiations. In **radiotherapy** it has been used to destroy cancer cells in the body, though X-rays which are not nuclear radiations are also used.

In the **food packaging** industry γ radiation can be applied to sealed bags of food in order to ensure that bacteria inside the bag are destroyed. This means that the shelf-life of the bag's contents is extended.

γ radiation can be used in the **detection of leaks** in underground oil or gas pipelines. If a small quantity of radioactive material is pumped under pressure along a pipe, it may leak into the ground and remain there after the pipe has been cleaned. A detector above ground can be passed over the pipe to detect the radiation above the place where the leak occurred. This spot is marked. An emitter with half life of a few days will soon decay in the ground to safe levels before engineers are sent to the spot to uncover the pipe and repair the leak (see Fig. 29.4).

In **medicine** a widely used radio isotope is technetium 99. This is a γ emitter with a half life of 6 hours, so its effect in the body is comparatively short lived. Technetium 99 is usually contained in the compound sodium pertechnetate, $NaTcO_4$.

A solution containing sodium pertechnetate can be administered either by swallowing or by direct injection into a vein. If swallowed it is absorbed by the gut; if injected it is pumped round the body in the blood stream by the heart. Either way it is soon widely dispersed in the body.

Many organs in the body, the kidneys, stomach, salivary glands, sweat glands and

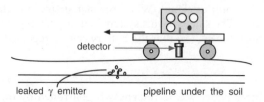

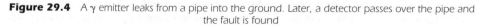

Figure 29.4 A γ emitter leaks from a pipe into the ground. Later, a detector passes over the pipe and the fault is found

others absorb the technetium-carrying ion, and they become radioactive as a result. By mounting a gamma sensitive camera alongside a patient's body, and studying how the radiation in a particular organ develops, it is possible to determine how well the organ is functioning. After a few hours the radiation from the absorbed ^{99}Tc isotope dies away to a negligibly low value.

Examining the functions of the brain

Fig. 29.5 shows a radiation picture formed by detecting gamma rays outside a person's head. The skull is seen from above. Researchers prepared a water sample with a small proportion of radioactive oxygen 15 in the water molecules. When a person drinks this water, their bloodstream becomes weakly radioactive. Concentrations of oxygen 15 atoms rise in areas where the rate of blood flow increases, and gamma detectors reveal these areas. Notice the light area indicating a particular region of the brain. There is increased radioactivity in this region. Just before the picture was taken, the person was asked to **think about** performing a particular finger movement, **but not actually to do it**. The picture shows the area of the brain involved in this thinking. Surprisingly, this part of the brain is not involved when a person is asked to carry out the finger movement.

As oxygen 15 has a very short half life (2.1 minutes), the blood of the person who drank the water is soon back to normal.

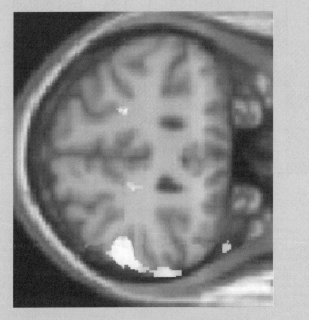

Figure 29.5 A brain scan showing γ ray emission from oxygen 15 nuclei (Wellcome Department of Cognitive Neurology, London)

Iodine 131 is another isotope used in medicine. It emits γ radiation with a half life of 8.1 days. Nearly all iodine in the human body is either excreted or absorbed in the thyroid gland. The speed of concentration of iodine in the thyroid gland, after a minute quantity of iodine 131 has been fed or injected into a patient, provides a measure of the thyroid function. Results are obtained with a γ-detecting 'camera' mounted close to the gland in the neck.

Medical students will come across a range of other techniques in which low levels of radiation may be used. For example surgical equipment could be freed of bacteria if exposed to γ radiation. Of course it is important that exposure to nuclear radiation does not make the exposed object radioactive.

Environmentalists and health inspectors find uses for nuclear tracers in studying the way sediment builds up in a bay or estuary. By adding some radioactive material to effluent pumped into a river, the flow of possible pollutants downstream may be predicted, and environmental damage prevented. It is also necessary to check coastal beaches near nuclear plants for low levels of radiation that should not be there.

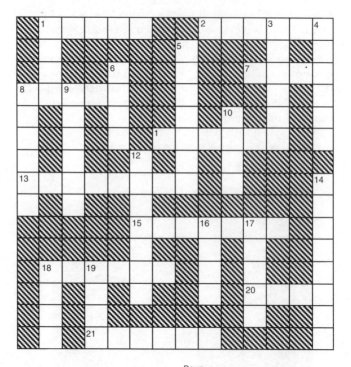

Clues

Across

1. A substance used to shut down a fission reactor. (5)
2. This number tells how many protons a nucleus contains. (6)
7. An object in the sky in which nuclear fusion occurs. (4)
8. A very penetrating nuclear radiation. (5)
11. A nuclear particle with positive electric charge. (6)
13. A charged particle, but not inside the nucleus. (8)
15 & 21. Some power stations derive electricity from this. (7, 6)
18. Atoms of this element can be formed from α radiation. (6)
20. If you cannot solve these clues, you may need this. (4)

Down
1. This nuclear radiation carries negative charge. (4)
3. A mass has kinetic energy when it is in this. (6)
4. This element is sometimes used in telling the age of organic remains. (6)
5. A particle which adds only mass to an atomic nucleus. (7)
6. This number tells how many protons plus neutrons a certain nucleus contains. (4)
8 & 9. They produced a tube to detect ionising radiations. (6, 6)
10. It consists of a nucleus surrounded by orbital electrons. (4)
12. It is contained in fuel rods for a nuclear fission reactor. (7)
14. Atoms of the same element, but having different masses. (8)
16. A good absorber of nuclear radiations. (4)
17. This radiation ionises air readily and soon loses its energy. (5)
18 & 19. You may need to know this in order to predict how the radioactivity of a source will decay. (4, 4)

Figure 29.6

Questions

1   What safety procedures are needed in running a nuclear power station?
2   Give an example of a useful application of nuclear radiation in an engineering context.
3   Give an example of the use of nuclear radiation in medicine.
4   Used fuel rods from a nuclear reactor have isotopes with long half lives. What problems arise from this?
5   List four things which add to the cost of using nuclear power for electricity generation.
6   Why is it short sighted to select an energy programme based entirely on fossil fuels?
7   Why is nuclear fusion not used for generating electricity?
8   What would be an advantage of a fusion reactor in commercial use?
9   Fig. 29.6 contains some revision questions. Tackle the crossword.

What have you learnt?

1   Do you know the sources of background radiation?
2   Can you give an example of when a nuclear source needs to have a long half life?
3   Do you know how nuclear radiation can effect living cells?
4   Do you understand the precautions needed in storing and handling radio-active materials?
5   Do you know the principles applied when:
   (a) engineers detect breaks in underground metal pipes?
   (b) production engineers control the thickness of material in a production process?
   (c) doctors study where things we eat are retained in organs of our bodies?
6   Do you know how the fission process is controlled in a nuclear reactor?
7   Can you explain the purpose of a moderator in a nuclear reactor?
8   Do you know how a fission reactor is shut down?
9   Can you give some reasons for adopting a pro-nuclear energy policy?
10   Can you give some reasons for rejecting a pro-nuclear energy policy?

Appendix A

A.1 Constant acceleration equations

The straight line graph in Fig. A.1 shows the motion of an object accelerating from velocity u to v in time t. The acceleration (a) is the gradient, because:

$$a = \frac{\text{gain in velocity}}{\text{time taken}} = (v - u)/t. \text{ From this } v = u + at \qquad (A.1)$$

The displacement (s), is the area under the graph. That means:

$$s = u \times t + \tfrac{1}{2}(v - u)t. \text{ From this we have: } s = \frac{(u + v)t}{2} \qquad (A.2)$$

Two other equations can be derived from these.

Eliminating v we get:

$$s = ut + \tfrac{1}{2}at^2 \qquad (A.3)$$

and eliminating t we get:

$$v^2 = u^2 + 2as \qquad (A.4)$$

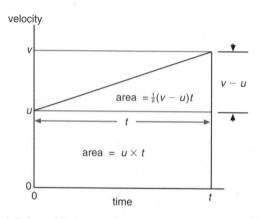

Figure A.1   Constant acceleration from velocity u to v in time t

These results only apply when the acceleration is constant, i.e. when the ratio of the resultant force to the accelerating mass is constant.

A.2 Momentum

Momentum is always a mass multiplied by its velocity. Newton's second law of motion relates the acceleration (a) of a mass (m) to the size of the resultant force (F) applied to it. Force equals mass $\times$ acceleration can be written:

$$F = m \times a \tag{A.5}$$

From Fig. A.1 and Equation A.1, acceleration can be written as:

$$a = (v - u)/t.$$

This means we can write:

$$F = m(v - u)/t \tag{A.6}$$

from which it follows that force $\times$ time = mass $\times$ velocity gain, i.e.

$$F \times t = (m \times v) - (m \times u) \tag{A.7}$$

or **impulse ($F \times t$) equals gain in momentum**. In collisions between two freely moving masses there is no externally applied force. Each mass exerts a force F, on the other for the same time t, and so each experiences an impulse of the same size. However, the impulses act in opposite directions, causing momentum changes in opposite directions. Hence the net change in the total momentum of the two masses is zero, since one loses momentum while the other mass gains an equal amount.

This is summed up in the **law of conservation of momentum**. For two masses m_1 and m_2 in collision while moving freely along the same line, we can write the equation:

initial total momentum = final total momentum, or,

$$m_1 u_1 + m_2 u_2 = m_1 v_1 + m_2 v_2 \tag{A.8}$$

The symbol u refers to initial velocities, and v refers to the final velocities of particles 1 and 2 after their collision.

A.3 Kinetic energy

Work and energy are measured in the same units (joules). When a resultant applied force does some work on a mass initially at rest, the mass moves as work is transformed into kinetic energy of the mass. The work done by the force is the force $\times$ distance it moves the mass. i.e. $W = F \times s$; and from Newton's equation of motion, $F = m \times a$, we can write the work done as:

$$W = m \times a \times s \tag{A.11}$$

In words this is saying that the **work done** by the resultant force is equal to the **gain in energy** of the mass. This energy, which speeds up its motion, is kinetic energy.

Now have a look at equation A.4 above. From this we can say that:

$$a \times s = (v^2 - u^2)/2.$$

We can rewrite the equation for work done as:

$$W = m \times a \times s = (1/2)m\ (v^2 - u^2) \tag{A.12}$$

For a mass initially at rest the initial velocity (u) is zero, and we can see that the **work done = gain in KE**, so that:

$$\textbf{the KE of the moving mass} = (1/2)m \times v^2 \tag{A.13}$$

This is a general result for the KE of any moving mass.

Appendix B Using mathematical equations in physics

B.1 Using equations in physics

Some physical changes may be described briefly with the use of equations to express the relationships between the variable quantities. Where letters are used, they represent the sizes of physical quantities, as for example P for pressure, F for force, m for mass, etc. Here are four types of equation that say things you might express just as well with words.

1 Some **quantities may simply be equal to others**, as for example when a stone falls freely and we write: **loss of PE of stone = gain in KE**. The quantities on each side of the equation are measured in the same units. Here, for example, the units are joules.

2 When some quantities change, their sizes may remain in a simple ratio, i.e. **they are in direct proportion to one another**. Their ratio can be expressed with a constant of proportionality. For example in a spring obeying Hooke's law, the extension x, is in proportion to the elongating force F. So the ratio $F/x = k$, or $F = k \cdot x$, where k is the constant of proportionality called the **stiffness** of the spring. The expression only applies over a limited range of force values, and you should always be aware of the conditions in which a mathematical equation can be applied to a physical situation. There are many examples of direct proportionality in physics.

3 Some quantities change so that **they remain inversely proportional** to one another. This means that if you double one quantity, the other is halved, *not* doubled. For example sound waves in air travel at a given speed, v. So v is a constant in the equation $v = f \times \lambda$. Here f is the frequency of the waves whose wavelength is λ. You will see that if f is doubled, then λ must be halved so that the product $f \times \lambda$ remains constant. Another way of writing the same result is $f = v/\lambda$, and λ **is said to be inversely proportional to** f.

4 The radioactivity of a sample **changes exponentially with time**. This means that, for equal additions of time, the activity is changed by the same factor. Suppose you know that the activity is reduced to 0.8 times its starting value in one day. Then after the next day the activity will have reduced again by a factor 0.8. After three days the activity will be only $0.8 \times 0.8 \times 0.8$ (i.e. it reduces to about 0.51 or half) of what it was initially. So we reckon the material has a half life of just over three days, in fact about 73 hours.

B.2 Some useful equations

1 Density = mass/volume. $\qquad\qquad\qquad\qquad\qquad$ $D = m/V$
2 Weight = mass × gravitational field strength. $\qquad$ $W = m \times g$
3 Work done = force × distance moved in the direction of the force. $\qquad\qquad\qquad\qquad\qquad\qquad\qquad$ $W = F \times s$
4 Increase of gravitational potential energy of mass m raised through height h = gain in PE = $m \times g \times h$
5 Power = energy transformed/time taken = work done/time taken.
$$P = W/t$$
6 Efficiency of a machine or system = $\dfrac{\text{useful energy output}}{\text{total energy input}}$

$\qquad\qquad\qquad\qquad\qquad\qquad = \dfrac{\text{useful power output}}{\text{total power input}}$

7 Moment of a force about a pivot = force × (perpendicular distance from the line of action of the force to the pivot).
8 In rotational equilibrium, (sum of clockwise moments about any point in their plane) = (sum of anticlockwise moments about that point).
9 Pressure = force/area. $\qquad\qquad\qquad\qquad\qquad$ $P = F/A$
10 For a fixed amount of an ideal gas, $\qquad$ $\dfrac{P_1 \times V_1}{T_1} = \dfrac{P_2 \times V_2}{T_2}$
11 Speed = distance/time taken. $\qquad\qquad\qquad$ $v = s/t$
12 Acceleration = change in velocity/time taken. $\quad$ $a = (v - u)/t$
13 Constant acceleration equations (see Appendix A).
14 Force = mass × acceleration. $\qquad\qquad\qquad\quad$ $F = m \times a$
15 Kinetic energy of mass m moving at velocity v. $\quad$ $\mathrm{KE} = \frac{1}{2}m \times v^2$
16 Momentum = mass × velocity. $\qquad\qquad\qquad$ $p = m \times v$
17 Impulse = resultant force × time = change in momentum. $F \times t$ = change in p
18 Frequency = 1/periodic time. $\qquad\qquad\qquad$ $f = 1/t$
19 Wave speed = frequency × wavelength. $\qquad$ $v = f \times \lambda$
20 Refractive index (n) = $\dfrac{\text{sine of angle of incidence}}{\text{sine of angle of refraction}} = \dfrac{\text{speed of light in medium 1}}{\text{speed of light in medium 2}}$

$\qquad$ or $n = \dfrac{\sin i}{\sin r} = \dfrac{v_1}{v_2}$

21 Energy supplied to raise temperature = mass × specific heat capacity × temperature change.
$\qquad\qquad$ or Q (joules) = $m \times c(T_2 - T_1)$
22 Charge = current × time. $\qquad\qquad\qquad$ Q (coulombs) = $I \times t$
23 Electrical energy = potential difference × charge. $\quad$ $W = V \times Q = V \times I \times t$
24 Electrical energy (in kWh) = power (in kW) × time (in h). $W = P \times t$
25 Electrical power = potential difference × current. $\quad$ $P = V \times I$
26 Potential difference = current × resistance. $\qquad$ $V = I \times R$
27 Electrical power released in a resistor. $\qquad\quad$ $P = I^2 \times R = V^2/R$

28 For a transformer with 100% efficiency:
$$\frac{\text{secondary voltage}}{\text{primary voltage}} = \frac{\text{number of secondary turns}}{\text{number of primary turns}} \quad \text{or} \quad \frac{V_s}{V_p} = \frac{N_s}{N_p}$$

29 For a transformer with 100% efficiency: secondary power output = primary power input or $V_s \times I_s = V_p \times I_p$

30 Truth tables for two input logic gates.

Inputs		Output			
		AND	OR	NAND	NOR
0	0	0	0	1	1
0	1	0	1	1	0
1	0	0	1	1	0
1	1	1	1	0	0

Appendix C

C 1. Revision questions from GCSE papers

These questions have been set in GCSE papers in the UK. Numbers in brackets show the number of marks available for each part of a question. We have shown where spaces were left for your answers, but our spaces may not be sufficient for the full answers that examiners hope to see.

A. The circuit diagram Fig. A.1 shows components (**L, M** and **N**) in series with a variable resistor.

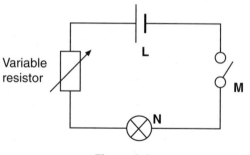

Figure A.1

(a) What are the names of components **L, M,** and **N?** _____(3)

(b) What do an ammeter and voltmeter measure?

    Ammeter _____Voltmeter _____(2)

(c) A student wanted to measure the voltage across **N** and the current in it. Draw the circuit diagram, and show how the voltmeter and ammeter should be connected. (2)

(d) The resistance of the variable resistor is increased. What effect does this have on:

    (i)   the current in **N:** _____(1)

    (ii) the voltage across **N?**_____(1)

(e) Some students thought that the current is used up as it goes around the circuit.

(i) How should they test to see if this is true? _____(2)

(ii) What would you expect them to find? _____(1)

(f) The diagram Fig. A.2 shows the electrical connections to a clothes iron.

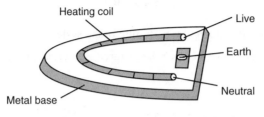

Figure A.2 *Electrical connections to a clothes iron*

(i) What is the colour of the insulation on the Earth wire?_____(1)

(ii) Explain the purpose of the Earth connection to the clothes iron. _____(1)

(iii) Which wire is connected through a **fuse**? _____(1)

(iv) Explain why the heating coil has to be **insulated** from the metal base of the clothes iron. _____(2)

(Total 17 marks) (London)

B. When car drivers see an emergency in front of them, there is always a short reaction time before they apply the brakes. The distance travelled during this delay is called the 'thinking distance'.

(a) (i) A driver moving at 20 m/s on a dry road takes 0.6 s to react. What is his thinking distance?_____(2)

(ii) At this speed the car's braking distance is 30 m. Find the TOTAL stopping distance for the car._____(1)

(iii) Driving conditions change. The roads may be wet. The driver may be tired. Driving conditions may change the thinking and braking distances. Complete the table stating whether the thinking and braking distances will increase, decrease, or remain the same.

Conditions	Thinking distance	Braking distance
Wet roads		
Tired driver		

(4)

(iv) Add to diagram Fig. B1 the direction of the braking force acting on the car. (1)

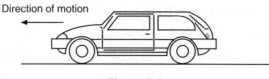

Direction of motion

Figure B.1

(b) The diagram Fig. B2 shows the hydraulic brake system of a car. When the driver's foot pushes on the brake pedal, a large force is exerted on the discs by the brake pads, and the car stops.

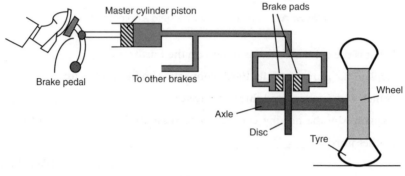

Master cylinder piston

Brake pads

Brake pedal

To other brakes

Wheel

Axle

Disc

Tyre

Figure B.2   (Not to scale)

(i)   Explain how a force is exerted by a brake pad when the brake pedal is pushed. _____(3)

(ii)   It is normal to use brake fluid (an oil) in the pipes to the brake pads. Explain why air would **not** be suitable._____(1)

(iii) The area of the master cylinder piston is 3.0 cm². The foot pushes with a force of 120 N. Calculate the pressure in the master cylinder._____(2)

(iv)   State the pressure exerted by the brake fluid on the brake pads. _____(1)

(v)   The area of each brake pad is 25 cm². Find the force exerted on each brake pad when the foot pedal is pushed._____(2)

(vi)   What energy transfer takes place when a car is stopped by its brakes? _____
_____(2)

(Total 19 marks) (London)

C. The graph (Fig. C1) shows how the drag force on a motor boat varies with its speed.

(a) (i)   Use the graph to find the value of the drag force acting on the boat when it travels at 1.5 m/s._____(1)

(ii)   What driving force is needed to keep the boat moving at a constant 1.5 m/s?

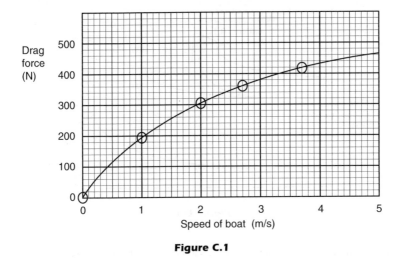

Drag force (N)

Speed of boat (m/s)

Figure C.1

Explain your answer._____(2)

(iii) How far does the boat move in 20 s when its speed is 1.5 m/s? _____(1)

(iv) Calculate the minimum work done by the engine in maintaining a speed of 1.5 m/s for 20 s. _____(2)

(v) Calculate the minimum power output of the boat engine during this time.

_____(2)

(vi) Explain why the engine's power is greater than the value found in part (v).

_____(2)

(b) When the boat is fully loaded with cargo and passengers the drag force at 1.5m/s is increased. Explain why the force is bigger._____(2)

(c) Cargo is normally placed below the water line. Explain why this is done.

_____(2)

(Total 14 marks) (London)

D. The diagram (Fig. D1) represents the energy changes taking place in a torch during one second.

(a) Use the information shown on the diagram to calculate:

(i) the total energy available from the battery during one second._____(1)

(ii) the efficiency of the lamp. _____(2)

(b) Describe **three** ways in which the thermal energy from the filament of the lamp is transferred to the surroundings _____(3)

(c) A pupil was asked to investigate how the electrical resistance of the lamp filament changed with the potential difference across it.

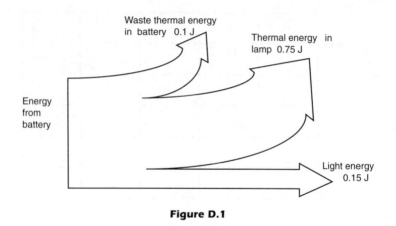

Figure D.1

(i)   Draw the circuit diagram for the arrangement she could use.   (3)

(ii) The data obtained from the investigation are given below.

Potential difference (V)	0	2.0	4.0	6.0	8.0	10.0
Current (A)	0	0.75	1.20	1.50	1.75	1.90

Plot a line graph of these results.   (3)

(iii) As the potential difference across the filament is increased, the temperature of the filament increases. What happens to the resistance of the filament as the potential difference increases? How does the graph show this?

_____(2)

(iv) The graph (Fig. D2) shows how the current in a filament lamp changes after it

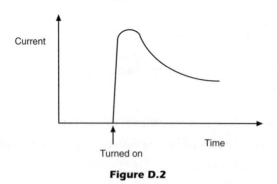

Figure D.2

is turned on. Use your answer to part (c) (iii) to explain why it behaves in this way. _____(2)

(Total 16 marks) (London)

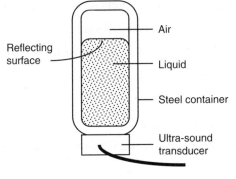

Figure E.1

E. Ultra-sound is used in industry for measuring the levels of liquids in containers. The Fig. E.1 shows the arrangement used. The transducer produces pulses of ultra-sound which pass through the liquid under test. By timing reflected pulses it is possible to calculate the position of the reflecting surface. In the liquid, the ultra-sound has a frequency of 10 MHz and speed of 1500 m/s.

(a) (i) Explain what is meant by ultra-sound. _____(2)

   (ii) Calculate the wavelength of the ultra-sound in the liquid. _____(3)

   (iii) Find how long it takes the pulse to travel from the transducer to the liquid's surface and back, when the liquid is 0.75 m deep. _____(4)

   (iv) Give two reasons why the reflected pulse will have a smaller amplitude than the transmitted pulse. _____(2)

(b) The transmitted and reflected pulses trigger an AND gate at X, so that the time interval between the two pulses is measured by the counting circuit shown in Fig.

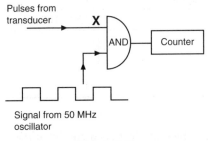

Figure E.2

E.2. The input X of the AND gate becomes high when the pulse is transmitted and becomes low when the reflected pulse is received.

(i) Explain why the signal from the 50 MHz oscillator is let through to the counter between the transmission and receipt of pulses at the transducer.__(1)

(ii) In one measurement, a count of 2000 pulses was recorded. Calculate the time between the transmitted and received signals. _____(3)

(iii) If the speed of the ultra-sound in the liquid was 1500 m/s, find the depth of the liquid in the container. _____(3)

(Total 18 marks) (London)

F. (a) Uranium -235 can be written $^{235}_{92}$U.

How many protons, neutrons, and electrons are in this **nucleus** of uranium?

_____(3)

(b) Uranium -235 is radioactive. It decays by alpha (α) emission to give thorium. The chemical symbol for thorium is Th.

(i) What is an alpha particle?_____(2)

(ii) Complete the equation below to show the decay of uranium -235 by writing the appropriate numbers in the boxes.

$$^{235}_{92}\text{U} \longrightarrow \boxed{}_{\boxed{}} \text{Th} + \boxed{}_{\boxed{}} \alpha \qquad (4)$$

(c) One form of carbon, called carbon -14, decays by beta (β) emission.

(i) What is a beta particle?_____(2)

(ii) Carbon -14 has a half-life of 5600 years. What does this mean?

_____(2)

(iii) Complete the table below to show the properties of alpha (α), beta (β) and gamma (γ) radiations. Some entries in the table have been completed for you.

	Alpha	Beta	Gamma
Range in air		15 to 100 cm	many metres
Charge compared to the proton	+2		
Mass compared to the proton		1/1840	

(5)

(d) In the carbon of a tree which has just been cut down there are 16 radioactive disintegrations per minute for **each gram** of carbon.

A wooden object from an archaeological site contains 8 grams of carbon, and it is decaying at a rate of 8 disintegrations per minute. The half-life of the radioactive carbon is 5600 years. Calculate the age of the wooden object. **Show clearly how you obtain your answer.** _____(6)

Sun   Earth

Figure F.1   (Not to scale)

(e) (i)   Which astronomical body is mainly responsible for tides on Earth?

_____(1)

(ii)   Name the form of attraction between the Earth and this astronomical body

causing the tides. _____(1)

(iii) Describe how spring tides are different from normal tides.

_____(2)

(iv) Show on Fig. F.1 the position of the Moon during a spring tide.   (2)

(f) The Sun (Fig. F.2) has been the same size for millions of years. This is because the

force which tends to increase the Sun's size is balanced by another force tending to

decrease it.

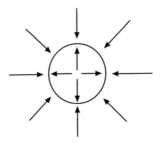

Figure F.2   Forces on the Sun

(i)   By what force are parts of the Sun pulled closer together? _____(1)

(ii) Name the process responsible for the Sun giving out energy._____(1)

(Total 32 marks) (Northern Ireland)

G. (a)   Two coils of insulated copper wire arranged opposite each other are shown in

Fig. G.1. One coil is connected to a cell through a switch. The second coil is

connected to a sensitive centre-reading ammeter.

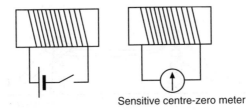

Sensitive centre-zero meter

Figure G.1

(i) Describe what is indicated by the ammeter when the switch is closed, then remains closed for ten seconds, and is then opened.

Switch is first closed. _____

Switch remains closed for 10s. _____

Switch is opened._____(4)

(ii) State three changes in the apparatus that would strengthen the observed effects.

1. _____

2. _____

3. _____(3)

(iii) What difference would it make to the observation if the cell were replaced with an alternating power supply producing a voltage of frequency 1 Hz, and the switch turned on? _____(2)

(b) Fig. G.2 shows part of a simple electricity generator.

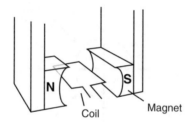

Figure G.2   A coil of insulated wire between unlike magnetic poles

(i)   What needs to be done to produce a voltage with this apparatus?_____(2)

(ii) Explain how this action generates an electricity supply with this apparatus. (3)

(c) Transformers are used to provide the right conditions for the efficient transmission of electricity over long distances. Fig. G.3 shows part of the transmission system. *T* is a transformer.

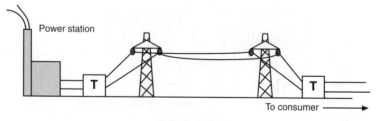

Figure G.3

(i) What name is given to the transformer used at the power station? Describe what it does, and why it is needed. Name: _____

What it does: _____

Why it is needed: _____(4)

(ii) What name is given to the transformer used where the power lines reach the consumer? Describe what it does, and why it is needed. Name: _____

What it does: _____

Why it is needed: _____(4)

(iii) What type of electricity, a.c. or d.c., is used by these transformers? _____(2)

(d) Hydroelectric and coal-fired are two types of large scale electricity generator. Compare these two types by completing the table below.

	Hydroelectric	Coal fired
Setting up costs	High	High
Running costs (high or low)	....	....
Environmental benefits	....	
Environmental drawbacks	....	....
Economic benefits		....

(6)

(Total 30 Marks) (Northern Ireland)

H. (a) Infra-red, ultra-violet and X-rays are electromagnetic waves.

(i) State **one** other thing they have in common and one difference. _____(2)

(ii) For each of the three radiations above, give one use it has in everyday life, and give one effect it has on the human body. Complete the table below on your answer sheet.

The use of the wave and its effect on the human body must be different. _____(6)

Type of wave	Use of the wave	Effect on the human body
Infra-red		
Ultraviolet		
X-rays		

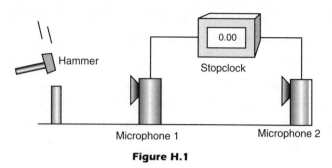

Figure H.1

(b) A method of measuring the speed of sound in air involves using two microphones and an electronic stop clock. Fig. H.1 shows how the apparatus is arranged. A sound is made by hitting a metal plate with a hammer. The distance between the two microphones is measured.

   (i)   What is the function of microphone 1 and microphone 2?

_____(4)

The distance between the microphones is varied, and for each distance the time for the sound to travel between them is measured. Measurements are shown in the table below.

Distance between the microphones (m)	0.2	0.4	0.6	0.8	1.0	1.2
Time for sound to travel (ms)	0.6	1.2	1.8	2.4	3.0	3.6

1 ms = 0.001 s

   (ii)   Use these results to plot a graph of distance (Y-axis) against time (X-axis). Use squared paper for the graph. (7)

   (iii) Use the graph to find the speed of sound. **Show clearly how you obtain your answer.** (5)

(c) (i)   What is meant by polarisation of a transverse wave?

_____(3)

   (ii)   Light reflected from a flat sheet of paper is polarised. Describe carefully how you would use a sheet of Polaroid material to show that this is so. State clearly what observation would support the statement.

_____(5)

(Total 32 marks)(Northern Ireland)

I. The circuit diagram Fig. I.1 illustrates one way of varying the current flowing in a circuit.

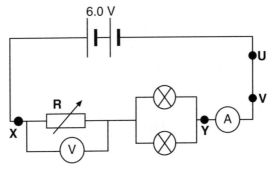

Figure I.1

The current is altered by changing the resistance of the variable resistor R. The battery voltage is 6.0 V, and the lamps are rated 3.5 V, 0.25 A. This means that when 3.5 V is applied to the bulb a current of 0.25 A flows in it, and the bulb lights normally. **Both bulbs are at normal brightness.**

(a) Calculate the **resistance** and the **power** of each lamp. **Show clearly how you obtain your answers.**

Resistance = _____(4)

Power = _____(4)

(b) What is the reading on the ammeter? _____(2)

(c) Calculate the **voltage** across the variable resistor R and its **resistance**. **Show clearly how you obtain your answers.**

Voltage = _____(2)

Resistance = _____(2)

(d) Calculate the **total resistance** between points X and Y. **Show clearly how you obtain your answer.** _____(6)

(e) The resistance of the variable resistor is now adjusted and the reading of the voltmeter increases. Will the lamps be brighter or dimmer than before? **Explain how you obtain your answer.** _____(4)

(f) Name the particles which move along the wires when an electric current is flowing round the circuit. On the circuit diagram mark, between U and V, the direction in which these particles move._____(2)

(g) Calculate the electric charge in coulombs that passes in **one minute**, through **one lamp** carrying a current of **0.25 A.**

Charge = _____C   (4)

(Total 30 marks) (Northern Ireland)

Figure J.1

J. Fig. J.1 is not to scale.

(a) (i) Add two arrows to the diagram to show the direction of:

(A) the pull of the Earth on the Moon (label this E);

(B) the pull of the Moon on the Earth (label this M). (2)

(ii) Explain why the Moon orbits the Earth. _____(4)

(b) Both the Sun and the Moon can be observed from the Earth's surface.

(i) What process occurs in the Sun to produce light? _____(1)

(ii) The Moon does not produce light of its own. How are we able to see the Moon? _____(2)

(c) The table below gives the surface temperatures of Mercury and Venus and their distances from the Sun.

	Maximum surface temperature (°C)	Distance from the Sun (millions of km)
Mercury	350	58
Venus	460	108

(i) Explain why the surface temperature of Mercury ought to be higher than that of Venus. _____(2)

(ii) It is thought that the atmosphere of Venus consists of a dense layer of carbon dioxide and sulphur dioxide. Explain in detail why this might cause the surface temperature of Venus to be higher than that of Mercury.

_____(3)

(d) Place the following in increasing order of size, starting with the smallest.

galaxy, Sun, universe, solar system, planet.

_____(1)

(e) Why would it be difficult for humans to travel to a distant galaxy?_____(3)

(Total 18 marks) (London)

K. (a) Fig. K.1 shows wavefronts of light in air arriving at a glass block.

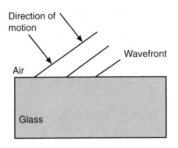

Figure K.1

(i) State what happens to the speed of the wavefront as it enters the glass block._____(1)

(ii) Complete the diagram to show the wavefronts once they have entered the glass. (1)

(iii) Explain what causes the wavefront to change direction as it enters the glass. _____(2)

(b) Fig. K.2 shows a simple camera which has been focused on a distant object.

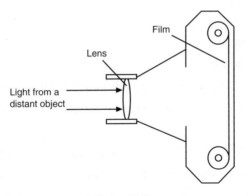

Figure K.2

(i) Continue the rays entering the lens to show how an image is produced on the film. (2)

(ii) What change is needed if the camera is to take a picture of a close object? Explain your answer._____(3)

(c) A camera is being used to photograph a passing car moving at a steady speed of 20 m/s. The shutter on the camera is opened for 0.05 s. The car is then 10m from the lens in the camera, and the film is 5.0 cm behind the lens.

(i) Calculate the distance the car travels during the time the shutter is open.

_____(2)

(ii) Explain how this causes a blurred image to be produced on the film.

_____(2)

(Total 13 marks) (London)

L. (a) A ray of light travels from air into glass as in Fig. L.1. The incident ray makes

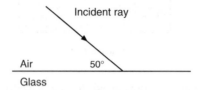

Incident ray

Air   50°

Glass

Figure L.1

an angle of 50° with the air-glass boundary. Some of the light is reflected at
the glass surface, but most passes into the glass and is refracted. The refracted
ray makes an angle of 60° with the air-glass boundary.

(i) Complete Fig. L.1 to show the reflected and refracted rays. Mark clearly the
angles of incidence, reflection and refraction.   (5)

(ii) Write the values of these angles in the correct spaces below.

Angle of incidence = _____

Angle of reflection = _____

Angle of refraction = _____(4)

(iii) Describe with the aid of a diagram how you could show that the image in
a plane mirror is the same distance behind the mirror as the object is in
front. _____(6)

Mirrors made from thick glass form multiple images of an object. Some of
these images are due to reflection at the glass surface, and some are due to
reflection at the silvered mirror surface. In Fig. L.2, O is a point object, I indicates an image.

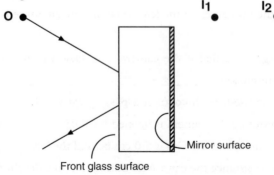

I_1 I_2

O

Mirror surface

Front glass surface

Figure L.2

(iv) Which image is due to reflection of light at the front glass surface?_____(1)

(v) Complete Fig. L.2 showing how rays of light are refracted and reflected so that a person sees the image I_2. _____(3)

(b) Oil companies surveying for oil set off explosions on the surface and study how the waves created by the explosion pass through the rocks, and are reflected at the boundary between rocks.

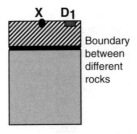

X   D₁

Boundary
between
different
rocks

Figure L.3

In Fig. L.3, X is the source of such an explosion, and D_1 is a detector of waves created by the explosion. Fig. L.4 shows the output from D_1, which only detects P type waves. When the P waves reach the detector, there is a sudden increase in the output from the detector.

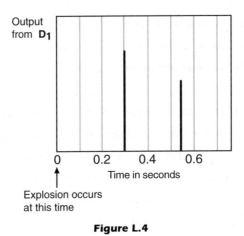

Output
from **D₁**

0   0.2   0.4   0.6

Time in seconds

Explosion occurs
at this time

Figure L.4

(i) How long after the explosion do the P waves **first** reach the detector?

_____(1)

(ii) P waves also reach D_1 at a later time. With the help of Fig. L.3 explain why this happens. The detector does not respond to sound waves travelling through air._____(2)

(iii) The P waves travel at a speed of 4 km/s through the rocks. Calculate **the distance along the surface, in km**, between the source X and the detector D_1. Show clearly how you obtain your answer.

Distance between X and D_1 = _____km (4)

The explosion at X also creates S waves. These travel at a speed of 2 km/s through the rocks. Detector D_2 detects only S waves (see Fig. L.5).

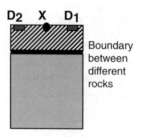

Figure L.5

(iv) On a grid like that in Fig. L.4, show how the output of detector D_2 will look when it detects S waves. **D_2 is the same distance from X as D_1.** (4)

(Total 30 marks) (Northern Ireland)

M. (a) Sam is carrying up some bricks on a building site. When lifted through a vertical distance of 2 m, **six** bricks gain 300 J of gravitational potential energy.

(i) Calculate the mass of a single brick. The acceleration, g, of free fall is 10 m/s². **Show clearly how you obtain your answer.**

Mass of a single brick _____kg (5)

To make it easier to raise bricks, Sam sets up a pulley system as shown in Fig. M.1. The load is 200 N and the tensions (T) in the rope are equal. The forces acting on the load are represented by the force diagram in Fig. M.1.

(ii) Calculate the **minimum** force which Sam must exert on the free end of the rope to hold it at rest. Assume there is no friction at the pulleys, and you can ignore the weight of the rope and pulleys.

Minimum force = _____N (2)

(iii) In practice, to raise the load at a steady speed, Sam must pull on the rope with a greater force than that calculated in part (ii). Give a reason for this. _____(1)

(b) A car jack is a force multiplier. It enables the user to produce a force large enough to lift a car (the load) by applying a small force (the effort). Fig. M.2 shows the

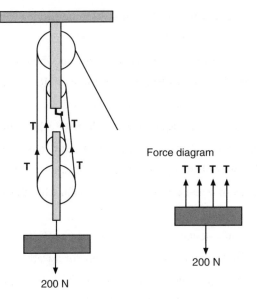

Force diagram

200 N

Figure M.1

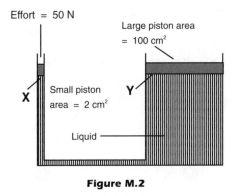

Effort = 50 N

Large piston area
= 100 cm²

Small piston
area = 2 cm²

X

Y

Liquid

Figure M.2

principle of the hydraulic jack. The effort is applied to the small piston, enabling the load (the car) to be lifted by the large piston.

(i)   A force of 50 N is applied to the small piston. Calculate the pressure in N/cm², produced in the liquid at X. **Show clearly how you obtain your answer.**

Pressure = _____ N/cm². (3)

(ii) Is the pressure in the liquid at Y greater than, equal to, or less than the pressure in the liquid at X? Give a reason for your answer. _____

Reason: _____ (2)

(iii) Calculate the upward force acting on the large piston at Y. **Show clearly how**

you obtain your answer. You may assume friction forces are so small they can be ignored. Force at Y = _____N. (3)

(iv) Calculate the work done on the small piston when it is moved down **5 cm**. **Show clearly how you obtain your answer**. Remember that the force acting downwards on the piston is 50 N. Work done _____(4)

(v) Calculate how far the liquid level at Y rises when the small piston moves down **5 cm.** The liquid used in the hydraulic jack is incompressible. **Show clearly how you obtain your answer.** Distance risen = _____cm. (3)

(vi) The efficiency of a real car jack is 0.8. Explain what this means. _____
_____(2)

(vii) In part (iii) you calculated the upward force at Y assuming that friction forces are negligible. In a real car jack friction forces are not negligible. Use your answers to parts (iv) and (v) to calculate the actual upward force at Y when the effort at X is 50 N. **Show clearly how you obtain your answer.** Remember the efficiency of the real car jack is 0.8.

Upward force at Y = _____N. (5)

(Total 30 marks) (Northern Ireland)

N. The diagram in Fig. N.1 shows the path of a ball when thrown through the air.

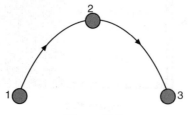

Figure N.1

(a) (i) Add to the diagram the names and directions of the force or forces acting on the ball in the three positions shown. Assume that position 1 corresponds to the point immediately after release and that air resistance is negligible. (2)

(ii) Explain, in terms of work done, why the ball reaches a maximum height.
_____(2)

(b) If the ball had been thrown with the same velocity on the Moon, describe and explain how the path of the ball would be changed. _____(4)
_____
_____

(Total 8 marks) (London)

Appendix D

D.1 Answers to numerical questions at the end of each chapter

Chapter 2:
2. $k = 0.5$ N/cm or 50 N/m, 3. 0.16 J.

Chapter 3:
7(b) Final pressure $= 3 \times$ starting pressure, 8(a) 300 K, 350 K, (b) $V_2 = (7\ V_1)/6$.

Chapter 4:
1. 10 mph, 3. 3.75×10^8 m, or 3.75×10^5 km, 4(a) 0.5 lengths, (b) zero, (c) 0.5 lengths, 5(a) 75 m, (b) 100 m, (c) 125 m, 6. 6.7 m/s, 9(a) 10 m apart, (b) 20 m apart, 10. 5 m/s/s, or 5 m.s^{-2}.

Chapter 5:
1. $F_R = 20$ N at about 37° with the 16 N force, 6(a) 2400 N, (b) -2400 N, 7(b) 2.4 m/s^2, (c) 6.3×10^5 N, (e) roughly 1.3 km, 10. 100 N.

Chapter 6:
4. 78.4 MJ, 5. 800 W.

Chapter 7:
6. 750 N, 7(a) 80 kg, (b) 128 N, 8(a) 1.6 m/s^2, (b) 1.6 m/s^2.

Chapter 9:
1. 5 N, 2(a) 60 J, (b) 80 J, (c) 75%, 3(d) 30 N, 4. 8×10^{-3} m^2, or 80 cm^2.

Chapter 10:
1. 500 s.

Chapter 11:
1. 1000 J/K, or J/°C, 6. 550 J/kg.K, 7. 168 kJ, 8. 0.44 kg.

Chapter 12:
4. 3×10^8 m/s.

Chapter 13:
3. 100 Hz, 6. 1300 m, 7(a) 10 ms, (b) (b), (c) (c).

Chapter 14:
11. 1024 Hz.

Chapter 15:
5. 1.5, 8(a) 5 minutes before 4 o'clock, (b) 25 minutes past 10 o'clock.

Chapter 16:
2. 5×10^{14} Hz, 3(a) 1.33×10^{-6} m, (b) 667 nm, 6. 22.95°.

Chapter 18:
4. 30 C, 5. 3 A, 6. 2.4 V, 7(a) 1.8 A, (b) away from the junction, 8(a) 24 C, (b) 120 J, 9. 1.5 V.

Chapter 19:

1(a) 50Ω, **(b)** 0.1 A, **(c)** 3 V & 2 V, **2(b)** 3 V, **3(b)** 0.1 A & 0.067 A, **(c)** 12Ω, **4(a)** 2, **(b)** 1, **(c)** 3, **5(a)** 4 A, **(b)** 10 A, **(c)** 60Ω, **(d)** 1.5 m, **7(a)** 200 V, **(b)** 100Ω, **(c)** 400 V, **(d)** 1600 J/s, **8(a)** 15Ω, **(b)** 20Ω, **(c)** 60Ω.

Chapter 24:

6(a) 1 A, **(b)** 100 A, **(c)** 1 W & 10 kW, **(d)** system (a) is more efficient.

Chapter 28:

7. 3.8 throws, **12(a)** 1 p 0 n, **(b)** 1 p 1 n, **(c)** 92 p 146 n, **(d)** 38 p 52 n, **13.** 88, 228.

Chapter 29:

9. Across: 1. boron, 2. atomic, 7. star, 8. gamma, 11. proton, 13. electron, 15 & 21. nuclear energy, 18. helium, 20. help.

Down: 1. beta, 3. motion, 4. carbon, 5. neutron, 6. mass, 8 & 9. Geiger Muller, 10. atom, 14. isotopes, 16. lead, 17. alpha, 18 & 19. half life.

Index

C

cable grip 207
camera 132, 156
  video 133
capacitance 236
capacitors 236–7
capacity
  thermal 97
car
  air bags 54
  brakes (*see* hydraulic)
  impact test 55
  versus runner 69
  seat belts 54
cathode ray 211–14
  deflecting 212
  oscilloscope (*see* oscilloscopes)
cell 171
centi- xvii
central heating 91–4, 93
centre of gravity 3
chain reaction 272
change of state 99–101
charge 168
  unit of 170
charging
  by rubbing 168
chain reaction 272
circuits 171–2, 174–5, 178–83, 193, 206,
    208, 226–32, 235–7
  lighting 208
  for stairs 208
circular motion 34–5
clockwise moments 46
coils
  primary 221
  secondary 221
  solenoid 193
collector 226–7
colour 132, 145, 155
  addition 155
  blindness 155
  filters 145
  primary 154–5
  secondary 155
  television tube 161
  triangle 155
  vision 154
comets 249–50
  Hale-Bopp 250
  Halley 250
  Shoemaker-Levy 250
communications satellites 65
commutator 199–202
compass 190
compression waves 110–11

Concorde 29,
condenser lens 155
conduction
  electrical 169–70
  thermal 87
conservation
  of energy (*see* energy)
  of momentum (*see* momentum)
constellations 253
convection 89
converging lenses (*see* lenses)
corkscrew rule 195
cost of electrical energy 209
coulomb 170
Crab nebula 257
critical angle 138–9
crystal structure 12–13
current 169–74
  alternating (*see* a.c.)
  direct (*see* d.c.)

D

dangers
  from high voltages 223
  of radioactivity 266
dark matter 259
d.c. 171, 204
  motors 199
  parts of 199
decay (*see* radioactivity)
deceleration 33–4
decibels 112, 126
demagnetising iron 194
density 2
  of air 3
detectors
  of light 60, 132, 179, 227–8
  of moisture 229–30
  of pressure 229
  of smoke 232
  of sound 112–13
  of temperature 21–2, 96, 179, 228
diffraction 107–8, 144–6
  grating 147
  of light 144–7
  of sound 117
  of water waves 107–8
diffusion 18
digital signals 224
diodes 181, 237
  light emitting 172, 181, 231
  rectifying 181, 237
direct current 171, 204
  motors 199
dispersion 138

P

Q

R